THE MOLECULAR GAZE

ART IN THE GENETIC AGE

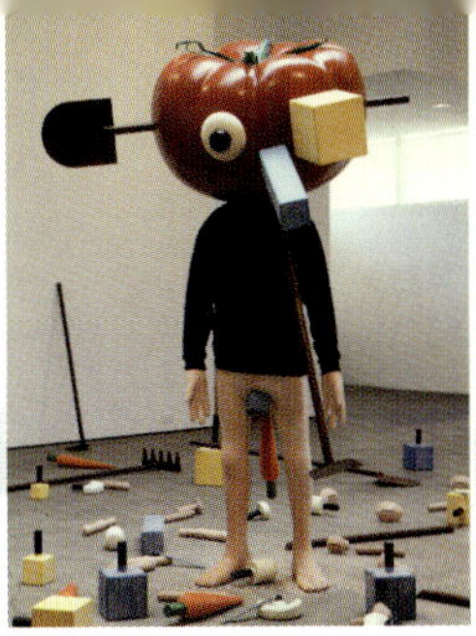

The Molecular Gaze
Art in the Genetic Age

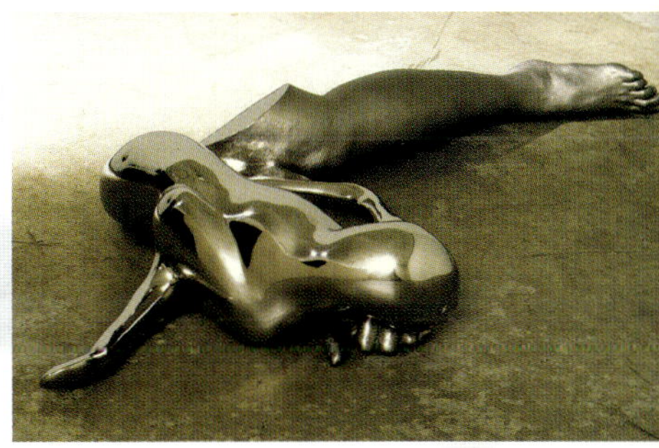

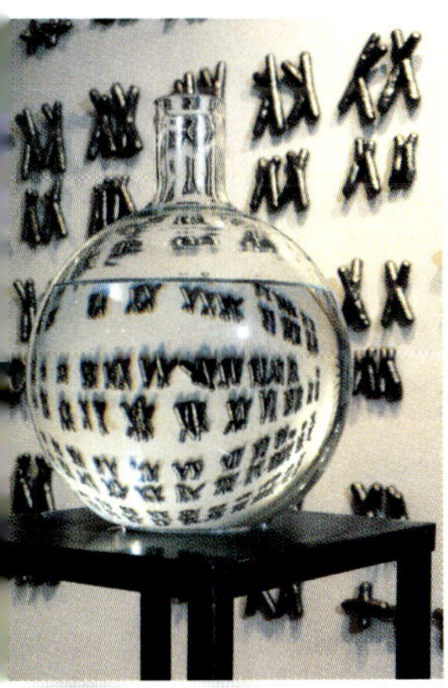

Suzanne Anker • Dorothy Nelkin
School of Visual Arts, New York *New York University*

Philip R. Reilly, series editor
The Cold Spring Harbor Laboratory Press Series on Genomics, Bioethics, and Public Policy

Cold Spring Harbor Laboratory Press
Cold Spring Harbor, New York

THE MOLECULAR GAZE
ART IN THE GENETIC AGE
The Cold Spring Harbor Laboratory Press Series on Genomics, Bioethics, and Public Policy

Printed in the United States of America

Publisher	John Inglis
Series Editor	Philip R. Reilly
Acquisition Editor	Judy Cuddihy
Developmental Editor	Judy Cuddihy
Project Coordinator	Mary Cozza
Permissions Coordinator	Nora Rice
Production Manager	Denise Weiss
Production Editor	Patricia Barker
Cover Designers	Suzanne Anker, Amy Wilson, Denise Weiss, Ed Atkeson

Front Cover:

Upper left: Paul McCarthy, *Tomato Heads* (detail), 1994. (Courtesy of the artist and Luhring Augustine, New York.)

Middle right: Rona Pondick, *Untitled Animal*, 1999–2000. (Courtesy of Sonnabend Gallery.)

Lower left: Suzanne Anker, *Zoosemiotics Primates, Frog, Gazelle, Fish* (detail), 1993. (Courtesy of the artist and Universal Concepts Unlimited, New York.)

Library of Congress Cataloging-in-Publication Data

Anker, Suzanne, 1946-
The molecular gaze : art in the genetic age / Suzanne Anker, Dorothy Nelkin.
p. cm.
Includes index.
ISBN 0-87969-697-4 (alk. paper)
1. Art and science. 2. Genetics in art. 3. Biology in art. 4. Science--Social aspects. I. Nelkin, Dorothy. II. Title.
N72.S3 A53 2003
701'.05--dc22

2003020067

10 9 8 7 6 5 4 3 2 1

All Cold Spring Harbor Laboratory Press publications may be ordered directly from Cold Spring Harbor Laboratory Press, 500 Sunnyside Blvd., Woodbury, N.Y. 11797-2924. Phone: 1-800-843-4388 in Continental U.S. and Canada. All other locations: (516) 422-4100. FAX: (516) 422-4097. E-mail: cshpress@cshl.edu. For a complete catalog of all Cold Spring Harbor Laboratory Press publications, visit our World Wide Web Site http://www.cshlpress.com/

To Erica

Dorothy Nelkin

July 30, 1933–May 28, 2003

Prolifically engaged and at home in the world, Dorothy Nelkin continuously rejected any assault on her free and independent will. *The Molecular Gaze,* her final book, on which she worked until the end of her days, is an awesome tribute to her fascination for uncovering explanations. Dot was passionately interested in contemporary art. She saw its presence as society's lens, sometimes bizarre, sometimes beautiful, sometimes provocative. Her insistence on taking on life as a creative and mutable journey survives in the enormous amount of work and love she leaves behind.

Contents

Preface

FROM DESIGNER BABIES TO ELIXIRS OF IMMORTALITY, the genetic age brings with it both fears and fantasies of science fiction. Restructuring evolution through technological intervention has been a continual process over time. Whether using live organisms to change the attributes of matter, as in beer brewing, or more recently in such novel technologies as nuclear transfer for cloning, the redirecting of nature by altering molecular structures continues. The year 2000 was a milestone in the scientific enterprise of genetics, bringing forth a rough draft of the human genome. Concomitantly, many exhibitions involving "art and genetics" began popping up internationally. This intersection between molecular biology and visual art is the subject of *The Molecular Gaze: Art in the Genetic Age.*

A collaboration between a visual artist and a sociologist of science, this book parallels developments in image-making and social policy during the 20th century. One of the authors (Anker) is a *Künstler*, whose work integrates molecular metaphors in paintings, sculpture, and installations. In addition, her fascination with visual images has led her into a potent engagement with art history. The other author (Nelkin) is a social scientist who for 30 years has written and researched in the area of science studies—namely, the significance of science for society. Over the past few decades, this work has come to focus on genetics—perhaps the field with the most profound influence on the way people deal with such basic issues as identity, life, birth, and family.

Both of us have been interested in the histories of art and culture. Both of us have written on the popular assimilation of genetic concepts—Anker from the perspective of the retinal, Nelkin from the use of verbal images in popular culture. This book was a natural outcome of this longstanding mutual interest. Making connections between the world as image and society as social space requires leaps of faith coupled with scrutiny. It is because of the diversity of our disciplines that this book balances pictorialization and interpretation with social policy.

We acknowledge the help of many people: Frank Gillette for his astute close reading and passionate support; Susan Lindee for joyfully reading the text in its entirety; Martin Kemp for his comments on an earlier article exploring the intersection of the semiotics of genetics with commodification; Judy Watkins and Vera Zolberg for a critical reading of parts of chapters; Amy Wilson for scanning, designing, refiguring, databasing, researching, and whatever else needed to be done; Gloria Houng for library and Internet research; Jocelyn Anker for constructing www.geneculture.org as an international avenue for the free flow of information; Kevin Redman for sending us some of the most extreme examples in this book; Mark Nelkin for technical assistance in converting incompatible computer files and code glitches; and Erica Peckens, Lisa Nelkin, and Mark Nelkin for nurturance in many ways.

We also thank Anne Edgar, Marvin Heiferman, Pamela Auchincloss, Aaron Panofsky, Carol Squiers, Howard Stein, Wolf-Dieter Stoeffelmeier, and Marian Ziola for their continued support of this subject; journalists and writers Sarah Boxer, Nancy Princenthal, Barbara Maria Stafford, Frances Terpak, Donna Haraway, Faye Hirsch, Heather Sparks, Steven Henry Madoff, Sian Ede, and Dan and Bettyann Kevles; Maryhelen Hendricks and Kevin Clarke for providing useful materials; and the College Art Association under the guidance of Ellen K. Levy for integrating this topic as part of their programming. We also thank the artists and galleries who have contributed their images and written material for this book; they are listed on page xvii.

At Cold Spring Harbor Laboratory Press, we are grateful to John Inglis, Judy Cuddihy, Liz Powers, Mary Cozza, Nora Rice, Denise Weiss, and Patricia Barker for their editorial efforts and for keeping this project on track. We are also grateful for the institutional support from the School of Visual Arts and New York University that has given us the opportunity to think creatively.

S.A.
D.N.

Divining DNA

If you were to stand on a busy street corner in Denver or Nashville or Atlanta and offer passersby a sawbuck to sketch the shape of any molecule, the vast majority (of those who could draw one) would quickly sketch a double helix, the iconic symbol of deoxyribonucleic acid. The aesthetically pleasing chemical structure of DNA was elucidated by James Watson, a brash young American biologist, and Francis Crick, a slightly more mature British physicist, at the Cavendish Laboratory in Cambridge, England in the spring of 1953. This discovery provoked perhaps the most understated comment in the history of science, when Watson and Crick wrote in their April 1953 *Nature* paper that "It has not escaped our notice that the specific pairing we have postulated immediately suggests a possible copying mechanism for the genetic material."[1] In contrast to this laconic remark, scientific folklore recounts that on the day on which the two men concluded that DNA was organized as a double helix, they rushed to a local pub where Crick loudly proclaimed, "We've discovered the secret of life!" He was more or less right.

Readers of *The Molecular Gaze*, an insightful exploration of the growing influence of DNA science on art, artists, and artistry during the last few decades by Suzanne Anker, herself an artist, and Dorothy Nelkin, a distinguished sociologist of science (recently deceased), will find a pleasing symmetry in the fact that the primordial representation of the molecule was a space-filling model several feet high made of brass struts and hand-tooled flat metal bases (representing the four nucleotides) produced by the laboratory's machine shop. That is, the double helix, with which artists today are so strongly bonded, was first offered to the world as a sculpture! Of even deeper aesthetic appeal is that the sole image of the molecule that appeared in the *Nature* paper which propelled Watson and Crick to scientific immortality and the Nobel Prize is a simple sketch drawn by Odile Crick, Francis's wife and, herself, an artist. Still more delightful for those who will embrace the themes of *The Molecular Gaze* is that the most enduring historical artifact of the discovery of DNA is a photograph made a month or so after the publication of the *Nature* paper. The photo captures the two molecular architects posed before their sculpture with Odile's sketch in the background, thus fusing three artistic mediums to present DNA to the larger world. A doubly autographed print of that photograph was auctioned off at an international meeting of geneticists in 2003 for a handsome sum. This is probably more money than the scientists expended in their 1953 research, and a splendidly ironic confirmation of the issues explored in a chapter in this book that some artists have recently raised concerning the commodification of genes.

In *The Molecular Gaze*, Anker and Nelkin assert that genetics is the text of choice for exploring the interplay of art with science in today's world. Just as Einstein's revision of space–time and the discovery of quantum physics induced a deep, sustained response in the art world in the 1920s and 1930s, over the last few decades Watson and Crick and the legions of creative scientists who are probing the molecular architecture of DNA have evoked extraordinary interest among artists across the globe. On reflection, there is really not a close contender to be the central text in the field dubbed sci-art. The poetry of space travel, caught in images of our blue and white lovely planet taken from the shuttle, has been marred by technical failures and tragic losses. The growing interest in biodiversity demonstrates the ascendance of ecology in the public mind, but (except for nature photography) few artists have made endangered species or loss of habitat a core theme in their work.

In the decade or so after the discovery of the double helix, it received relatively little attention, even in the scientific literature. In 1960, *Nature*, then as now a premier scientific journal, published nearly 50 papers concerning DNA, of which the double helix was mentioned in only two![2] This is probably because scientists had not developed the tools they needed to explore the consequences of the discovery of how all life is replicated. Here art paralleled science. In the first decade after the discovery of the double helix, only one major artist, Salvador Dali, explicitly represented the DNA structure, first in 1957 and again in 1963.

As I compose these words, I realize that the first time I seriously thought about DNA was as an undergraduate at Cornell when I contemplated *Galacidalacidesoxyribonucleicacid*, Dali's 1963 work. I recall (as the authors note) accompanying text, in which Dali mused that the double helix was the "only structure linking man to God." Several years later, in the early days of my circuitous journey into genetics, I sat one sultry July afternoon in the Olin Library at Cornell and read for the first time the 1953 *Nature* paper. That summer I was a scientific novice, working in Adrian Srb's yeast genetics lab. I mention this trivial point because Srb was both a great classical geneticist and deeply interested in art and architecture (especially, the great cathedrals). My memory of that afternoon in the library is crisp; for me both the molecule and the men who discovered its structure were already icons.

The fact that in my case exposure to an artist's vision of the double helix preceded the development of both my abiding interest in the scientific exploration of DNA and my deep interest in the ethical issues and societal controversies that flow from our ability to manipulate DNA molecules (a concern central to much of the artistic work examined by the authors) is only part of my strong sense of being bonded to this book. Consider in addition that it was I who asked Dorothy Nelkin that she and her co-author consider publishing the manuscript about art and science that became *The Molecular Gaze* with Cold Spring Harbor Laboratory Press, that Cold Spring Harbor is the laboratory which Watson led so effectively for so many years, and that my wife and I were influenced by Georges Seurat (the great French pointillist) in conceptualizing the cover for my most recent book about genetics, also published by this fine press. This confluence of coincidences makes me a bit giddy—like the slightly dizzy rush a child might feel sliding down the polished handrail of a spiral staircase.

Beginning in the 1970s, the field of molecular genetics acquired a series of ever more powerful tools (restriction endonucleases, the polymerase chain reaction) with which scientists could perform laboratory feats that in the 1960s were unimaginable. I can still feel the electricity in the air in a lecture room in Houston when in the mid 1970s a standing-room-only crowd of scientists gathered to hear Paul Berg, himself a Nobel laureate, lecture on the use of restriction enzymes to cut and paste DNA, in essence to reengineer living things. That day the talk was of moving genes around in viruses; today it is of moving human genes into goats; tomorrow it will be of enhancing the human genome. No wonder the artists have taken notice!

In exploring how molecular genetics has been integrated into visual culture, Anker and Nelkin chose five themes which they argue dominate the genre of "genomic art." These are the reduction of the body to molecular text, the meaning of mutation, the blurring of species boundaries, the notion of perfection, and the commodification of nature. Given the immense scientific emphasis that has been devoted to developing a consensus sequence of a human being (an achievement that was announced by two successive U.S. presidents), the idea that an unimaginably well-packed chemical thread more than 3,000,000,000 nucleotides long constitutes one-half of the bipartite software code (our program starts operating only with the fusion of sperm with egg) that is unique in each conception is a powerful image. Autoradiographs, the bar code pattern made by particles emitted by short sequences of radioactively labeled DNA, were among the first symbols of our molecular selves that artists transformed. *Designer Gene* (1992), a work by Dennis Ashbaugh, is a large painting that uses Rothko-like patterns of color to transfer the image of the radiograph into an artistic code. Those readers schooled in classical bacterial genetics may think of this as replica plating. Another artist, Kevin Clarke, approaches the reductionist concerns implicit in the concept of a human code by fashioning portraits out of the letters A,T, G, and C, the universal symbols for the four nucleotides. On February 15, 2001, *Nature* imitated art when it published a cover presenting the double helix as a montage of thousands of small human faces. The idea for the image, if not the thing itself, was developed by Eric Lander, one of the world's leading genome scientists.

In molecular biology, mutation is a neutral concept, used to note a change in some aspect of the DNA molecule; for example, the substitution of one of the four nucleotides by another. Mutations are essential to life. They are the stuff upon which the forces of natural selection work. Anker and Nelkin point out that artists have (albeit, until a few decades ago, unwittingly) long been fascinated with the consequences of deleterious mutations. For 2000 years poets and painters have created monstrosities. In the *Odyssey*, Homer imagined the Cyclops. From the Renaissance forward, artists have frequently depicted achondroplastic dwarves (individuals born with a mutation in a gene that curtails growth of the limb bones) to symbolize nature gone awry. Anker and Nelkin lead us to *Portrait of a Dwarf* (1987), in which photographer Joel-Peter Witkin poses a female dwarf dressed in white lace lingerie between an animal carcass and an idealized Greco-Roman bust, as though she were a rung on the 19th-century evolutionary ladder.

Transgenics, our ability to move genes across species barriers, has held great allure for artists. In this realm of modern biology, scientists are giving physical reality to more subtle versions of the chimeras, griffons, and centaurs that populate our foundational mythology. Artist Eduardo Kac has called himself the world's first transgenic artist. He has worked directly with geneticists to attempt to create animals that are genetically engineered to fluoresce (using the gene for a substance called green fluorescent protein). Three Australian artists have worked with cell biologist Dr. Joseph Vacanti on *The Pig Wings Project* (2001–2002), an effort to sculpt bird, bat, and reptile wings from mounds of pig cells. Anker and Nelkin show us that some artists are horrified by transgenics while others are liberated.

The notion that we might harness molecular biology to produce superior children has fascinated scientists and artists alike since the birth of classical genetics in the early years of the 20th century. For decades Aldous Huxley's novel, *Brave New World* (1932), which opens with a tour of a human hatchery in which extracorporeal wombs are bathed in special chemicals to shape the intellectual potential of future human beings, was a symbol of our urge to move human reproduction from genetic roulette to a highly engineered activity with a guaranteed product. Anker and Nelkin suggest that most modern artists who have confronted eugenics have a deep sense of foreboding. Many are made uneasy by this notion of engineering a self. Swiss artist Hans Danuser's *Frozen Embryo Series I* (1998–2000) consists of a series of minutely detailed photographs of human embryos placed against an abstract field. They remind Anker and Nelkin of frozen food, and prompt them to ask whether they have an "expiration date."

In their penultimate chapter, the authors guide the readers on a tour showing how artists have reacted to the rise of the biotechnology industry. They focus on the commodification of genes, the transformation of animals into biological tools, and the notion that our society now recognizes the right (under certain circumstances) to own DNA sequences that are the products of millions of years of evolution. Anker and Nelkin are particularly drawn to the oncomouse, an animal genetically engineered by Harvard scientists to gain more insight into the function of human breast cancer genes. The oncomouse is the world's first genetically engineered mammal to be protected by a patent (U.S. Patent No. 4,736,866). In Bryan Crockett's *Ecce Homo* (2000), a six-foot-tall, hairless oncomouse is depicted as "the ultimate figure of salvation." The oncomouse has replaced Christ, as science (the artist holds) has replaced religion.

The cover of an issue of *Time* commemorating the discovery of the double helix adumbrates *The Molecular Gaze* and recalls the early genetic meditations of Salvador Dali. In *Butterfly Landscape, The Great Masturbator in Surrealist Landscape with DNA* (1957–1958), Dali portrayed DNA as a tightly packed coil of blue marble like globules encasing golden globules of equal size from which were emerging a seemingly unending stream of butterflies. Dali's DNA was set against a surreal landscape and light blue sky. The *Time* cover depicts two sexually alluring humans, a white man and a biracial woman, each encased in a golden double helix that above them intertwines and morphs into a single tree (of life) floating rootless in a pale blue sky, itself a surreal landscape. This Daliesque cover presages the wonderful art tour that Anker and Nelkin provide their readers. Scientists try to solve the mysteries of DNA; artists contemplate them. *The Molecular Gaze* wonderfully illustrates how the two worlds are entwined.

Notes

1. Watson J.D. and Crick F.H.C. 1953. Molecular structure of nucleic acids: A structure for deoxyribose nucleic acid. *Nature* 171: 737–738.
2. Olby R. 2003. Quiet debut for the double helix. *Nature* 421: 402–405.

Philip R. Reilly, MD, CEO
Interleukin Genetics

Image Credits

Reductionism: The Body as a "Code Script" of Information

1. Marc Quinn, *Sir John Sulston: A Genomic Portrait,* 2001. Stainless steel, polycarbonate agar jelly, bacteria colonies, cloned human DNA. © The artist. From the collection of Jay Jopling/White Cube, London.
2. Franz Joseph Gall, Phrenology Diagram.
3. Charles le Brun (17th century). Top: *Etude de Taureau et de Boeuf.* Bottom: *Visage.* Courtesy of Paris Museé du Louvre, Department des Arts Graphique.

4a. Honoré-Victorin Daumier, *Jacques Lefebvre: The Fine, Trenchant Spirit (L'Esprit fin et Tranchant),* (ca. 1832–1835, collection Hirshhorn Museum). Courtesy of the Hirshhorn Museum and Sculpture Garden, Smithsonian Institution. Gift of Joseph H. Hirshhorn, 1966.

4b. Honoré-Victorin Daumier, *The Parisian Janitor (Le Portier Parisien),* (ca. 1840–1862, collection Hirshhorn Museum). Courtesy of the Hirshhorn Museum and Sculpture Garden, Smithsonian Institution. Gift of Joseph H. Hirshhorn, 1966.

5. Edgar Degas, *The Little Fourteen-year-old Dancer* (left profile), ca. 1880, cast 1922. All rights reserved, The Metropolitan Museum of Art, Bequest of Mrs. H.O. Havemeyer, 1922.(29.100.370).
6. Christian Boltanski (French, b. 1944), *The Detective* (detail), 1972–1973. 408 Photographs in various sizes, 1974.223/© 2003 Artists Rights Society (ARS), New York/ADAGP, Paris. Courtesy of The Art Institute of Chicago.
7. Juxtaposition of works of "degenerate" art by Karl Schmidt-Rottluff and Amedeo Modigliani and photographs of facial deformities. Reprinted from Schultze-Naumburg, Paul. 1928. *Kunst und Rasse.* J.F. Lehmann, München. Top row, left: Karl Schmidt-Rottluff; top row, second from left: Amedeo Modigliani; bottom row, left: Karl Schmidt-Rottluff, *Self-Portrait with Hat,* 1919. (Oil on canvas; Framed: 85 x 77 cm; Unframed: 73.3 x 65 cm. Collection: The Cleveland Museum of Art); bottom row, second from left: Karl Schmidt-Rottluff, *Portrait of Emy,* 1919. (Oil on canvas; 28 5/16 x 25 3/4 in. [73.5 x 65.5 cm]. Collection: North Carolina Museum of Art.)
8. Paul Gauguin (French, 1848–1903), *Tehamana Has Many Ancestors (Merahi metua no Tehamana),* 1893, oil on canvas, 76.3 x 54.3 cm. Gift of Mr. and Mrs. Charles Deering McCormick, 1980.613. Courtesy of The Art Institute of Chicago © 2001.
9. Henri Michaux, *Alphabet* (detail), 1927. © 2003 Artists Rights Society (ARS), New York/ADAGP, Paris. Private collection.
10. Henri Michaux, *Untitled (Movements [Mouvements]),* 1950–1951. © 2003 Artists Rights Society (ARS), New York/ADAGP, Paris. From a private collection.
11. Jackson Pollock, *Stenographic Figure,* 1942. © 2003. The Pollock-Krasner Foundation/Artists Rights Society (ARS), New York. Courtesy of The Museum of Modern Art, New York; Mr. and Mrs. Walter Bareiss Fund.

12. Tracy Wellman, *Kingsley Read's Shaw Alphabet* (after Pitman). Drawing by Tracy Wellman. From *The Story of Writing* by Andrew Robinson, 1995. Thames & Hudson, London and New York.
13. Salvador Dali, *Butterfly Landscape, The Great Masturbator in Surrealist Landscape with DNA*, 1957–1958. © 2003 Gala-Salvador Dali Foundation/Artists Rights Society (ARS), New York/Photo Descharnes/www.daliphoto.com. Private collection.
14. Yves Klein, *Anthropométrie de l'époque bleue (Anthropometry of the Blue Period)*, 1960. © 2003. Artists Rights Society (ARS), New York/ADAGP, Paris. From the collection of the Museé National d'Art Moderne Centre Georges Pompidou.
15. Zaire drawing, Mbuti tribal design (20th century). From *Mbuti design: Paintings by Pygmy Women of the Ituri Forest* by Georges Meurant and Robert Farris Thompson, 1996. Courtesy of Hansjorg Mayer, London.
16. Micrograph of sex chromosomes. *Cytogenetics of Mammalian X Chromosomes Part B*, edited by Avery A. Sandberg. © 1983 A.R. Liss. Reprinted by permission of Wiley-Liss, Inc., a subsidiary of John Wiley & Sons, Inc.
17. Conventionalized human figures from Spanish petroglyphs, compared with similar designs from the painted pebbles of Mas d'Azil, Ariege. After H. Obermaier. From Hugh Obermaier, *Fossil Man in Spain*. Published for the Hispanic Society of America by Yale University Press, 1924.
18. Karyotype of Hammer-headed Fruit Bat. Reprinted, with permission, from *An Atlas of Mammalian Chromosomes*, © 1970 Springer-Verlag.
19. Tony Cragg, *Code Noah*, 1988. Courtesy of the Marian Goodman Gallery. From the collection of Mr. and Mrs. Ware Travelstead.
20. Suzanne Anker, *Micro Glyph (Soma Font)*, 2000. From the collection of Goldblum Lévy, Switzerland.
21. Suzanne Anker, *Zoosemiotics: Primates, Frog, Gazelle, Fish* (detail), 1993. Courtesy of the artist and Universal Concepts Unlimited, New York.
22. Gary Schneider, *Chromosomes* (from *Genetic Self Portrait*), 1997. (Detail/Y chromosome.) Courtesy of the artist.
23. Dennis Ashbaugh, *Designer Gene*, 1992. Courtesy of the artist.
24. Iñigo Manglano-Ovalle, *Lu, Jack, and Carrie* (from *Garden of Delights*), 1998. Courtesy of Max Protetch Gallery.
25. Kevin Clarke, *Portrait of James D. Watson* (panel 5), 1998–1999. Courtesy of the artist.
26. Kevin Clarke, *Portrait of Jeff Koons*, 1993. Courtesy of the artist.
27a. Steve Miller, *Eat Protein #230*, 2002. Courtesy of the artist.
27b. Steve Miller, *Eat Protein #259*, 2002. Courtesy of the artist.
28. Michael Joaquin Grey and Randolph Huff, *Carbon Codex 15 x 8*, 1992. Graphite tablet, CNC milled, 61 x 122 x 5 cm. Photo by Stephen White, London. Courtesy of bitforms Gallery.
29. Olivia Parker, *Forms in Blue*, 1999. Iris Inkjet Print, 1999. Courtesy of the Robert Klein Gallery.
30. Zhang Huan, *Family Tree*, 2000. Courtesy of the artist.
31. Benjamin Fry, *Genomic Cartography: Chromosome 21*, 2001. Courtesy of the artist and the Aesthetics & Computation Group, MIT Media Laboratory, Cambridge.
32. Antony Gormley, *Bodies in Space IV*, 2001. Forged ball bearings, 206 x 86 x 72 cm. © The artist. Photo by Stephen White. Courtesy of Jay Jopling/White Cube, London.

Mutation, Manipulation, and Monsters: The New Grotesque in Art

1. Jake and Dinos Chapman, *Tragic Anatomies* (detail), 1996. © The artists. Courtesy of Jay Jopling/White Cube, London.
2. Diego Velázquez, *Las Meninas,* 1656. © Museo Nacional del Prado, Madrid.
3. Mr. Dent, Plates II and III. Two watercolor drawings of the Two-Headed Boy of Bengal, by a Mr. Dent. Used by Everard Home in his 1799 paper on the boy. Engraved for the Philosophical Transactions of the Royal Society. © The Royal Society. Taken from *Philosophical Transactions,* 1799, Vol. 89, pp. 28–30.
4. Bruce Nauman, *Ten Heads Circle/Up and Down,* 1990. © 2003 Bruce Nauman/Artists Rights Society (ARS), New York. Courtesy of the artist.

5a. Matthew Brady, *Barnum's Living Skeleton.*

5b. Matthew Brady, *Barnum's Fat Lady.*

6. Cornelia Hesse-Honegger, *Four Seed Bugs,* Lygaeus equestris, *from Tubre and Santa Maria Madonna, Italy,* 1994–1995. Courtesy of the artist.
7. Scott Lindgren, *Cephalothoracopagus,* 1990. Courtesy of the artist.
8. Odilon Redon, *The Cyclops,* ca. 1914. Courtesy of the artist. From the collection of the Museum Kroller-Mueller, Otterlo, The Netherlands.
9. Hannah Höch, *Strange Beauty* (*Fremde Schoenheit*), 1929. © 2003 Artists Rights Society (ARS), New York/VG Bild-Kunst, Bonn. From the collection of Berlinische Galerie, Landesmuseum für Moderne Kunst, Berlin.
10. René Magritte, *(L'Esprit de Géométrie) The Spirit of Geometry,* 1937. © 2003 C. Herscovici, Brussels/Artists Rights Society (ARS), New York. From the collection of the Tate, London.

11a. Hans Bellmer, *La Poupée (The Doll)* (detail), ca. 1937–1938. © 2003 Artists Rights Society (ARS), New York/ADAGP, Paris. From the collection of the Tate, London.

11b. Hans Bellmer, *La Poupée* (*The Doll*) (detail), 1937–1938. © 2003 Artists Rights Society (ARS), New York/ADAGP, Paris. From the collection of the Tate, London.

12. John Isaacs, *The Matrix of Amnesia,* 1997. Private collection.
13. Rona Pondick, *Untitled Animal,* 1999–2001. Courtesy of the Sonnabend Gallery.
14. Jake and Dinos Chapman, *Zygotic Acceleration, biogenetic, de-sublimated libidinal model (enlarged x 1000),* 1995. © The artists. Courtesy of Jay Jopling/White Cube, London.

15a. Cindy Sherman, *Untitled,* 1992. Courtesy of the artist and Metro Pictures.

15b. George Gessert, *Streptocarpus hybrid 523,* hybridized 1997, first bloomed 1998. Flowers 2" across. Courtesy of the artist.

16. Nancy Burson, *Untitled (Jessica and Her Mom),* 1991. Courtesy of the artist.
17. Rosemarie Trockel, *Hydrocephalus II,* 1982. © 2003 Artists Rights Society (ARS), New York/VG Bild Kinst Bonn. From the Joshua Gessel Collection.
18. Rosamond Purcell, *19th Century Skeleton of Hydrocephalic Child,* 1995. Photographs © R.W. Purcell 2000.
19. Joel-Peter Witkin, *Portrait of a Dwarf,* 1987. Courtesy of the artist. From the collection of the Fraenkel Gallery, San Francisco.
20. Orlan, *Orlan with Imaginary Generic no. 31: Successful Operation(s),* 1990. © 2003 Artists Rights Society (ARS)/ADAGP, Paris.
21. Michael Rees, *Ajna Spine Series 1.11,* 1998. Courtesy of the artist.
22. Charles Ray, *Family Romance,* 1993. Mixed media, 54 x 96 x 24 inches, Edition of three. Courtesy of Donald Young Gallery.

Blurring Boundaries: Chimeras and Transgenics

1. Thomas Grünfeld, *Misfit (St. Bernard),* 1994. Courtesy of the artist. From the Saatchi Collection, London.
2. Artist Unknown, Feejee mermaid. Peabody Museum, Harvard University, Photo T1383. © President and Fellows of Harvard College, 2003.
3. Rosemarie Trockel, *Untitled (Pennsylvania Station),* 1987. © 2003 Artists Rights Society (ARS), New York/VG Bild Kinst Bonn. From the collection of the Barbara Gladstone Gallery.
4. Max Ernst, *Stratified Rocks, Nature's Gift of Gneiss Lava,* 1920. © 2003 Artists Rights Society (ARS)/ADAGP, Paris. From the collection of the Museum of Modern Art, New York.
5. Max Ernst, *Here Everything is Still Floating (Le Vapeur et le Poisson),* 1920. © 2003 Artists Rights Society (ARS)/ADAGP, Paris. From the collection of the Museum of Modern Art, New York.
6. Hannah Höch, *Monument I,* 1924. © 2003 Artists Rights Society (ARS), New York/VG Bild Kinst Bonn. From the collection of Jean-Paul Kahn, Paris.
7. Raoul Hausmann, *Tatlin at Home,* 1920. © 2003 Artists Rights Society (ARS)/ADAGP, Paris.
8. Phosphorescent diagnostic cells. Reprinted with permission from Perry A.C., Wakayama T., Kishikawa H., Kasai T., Okabe M., Toyoda Y., and Yanagimachi R. 1999. Mammalian transgenesis by intracytoplasmic sperm injection. *Science* 284: 1180–1183. © 1999 American Association for the Advancement of Science.
9. David Ow, Tobacco seed combined with firefly genes. Reprinted, with permission, from Ow D.W., Wood K.V., DeLuca M., de Wet J.R., Helinski D.R., and Howell S.H. 1986. Transient and stable expression of the firefly luciferase gene in plant cells and transgenic plants. *Science* 234:856–859. © 1986 American Association for the Advancement of Science.
10. Glow-in-the-dark mice. Courtesy of Advanced Cell Technology.
11. Eduardo Kac, *GFP Bunny,* 2000. Alba, the fluorescent rabbit. Courtesy of Julia Friedman Gallery, Chicago.
12. Tissue Culture and Art Project (Oron Catts, Ionat Zurr, and Guy Ben-Ary), *Pig Wings Project: The Chiropteran Version, the Aves Version, and the Pterosaurs Version,* 2000–2001. Medium: Mesenchymal cells (bone marrow stem cells) from pig and biodegradable/bioabsorbable polymers. Courtesy of The Tissue Culture & Art Project.
13. Mouse genetically engineered with a human ear. Reprinted, with permission, from Cao Y., Vacanti J.P., Paige K.T., Upton J., and Vacanti C.A. 1997. Transplantation of chondrocytes utilizing a polymer-cell construct to produce tissue-engineered cartilage in the shape of a human ear. *Plastic Reconstructive Surgery* 100: 297–304.
14. Paul McCarthy, *Tomato Heads* (detail), 1994. Courtesy of the artist and Luhring Augustine, New York.
15. Jane Schneider, *What's That?,* 2002. Courtesy of the artist and June Kelly Gallery.
16. Matthew Barney, *CREMASTER* 4, 1994. Production Photograph. © 1994 Matthew Barney. Photo by Michael James O'Brein. Courtesy of Barbara Gladstone Gallery.
17. Gerhard Lang, *Palaeanthropical Physiognomy* (detail), 1991/2000. Courtesy of the artist.
18. Bradley Rubenstein, *Untitled (Boy with Puppy-dog Eyes),* 1994–1995. Courtesy of the artist and Universal Concepts Unlimited, New York.
19. Alexis Rockman, *Biosphere: Laboratory,* 1993. © 1993 Alexis Rockman. Courtesy of Gorney Bravin + Lee, New York.

20. Alexis Rockman, *The Farm,* 2000. © 2000 Alexis Rockman. Courtesy of Gorney Bravin + Lee, New York. From the Collection: The Joy of Giving Something, New York.
21. Frank Gillette, from the *HCE Series,* 1999–2000. Courtesy of the artist and Universal Concepts Unlimited, New York.
22. Frank Gillette, *A Tangled Bank,* 2000. Courtesy of the artist and Universal Concepts Unlimited, New York.
23. Eva Sutton, *Hybrids* (detail), 2000. Courtesy of the artist.
24a. Aziz + Cucher, *Chimera #1,* 1998. Color photograph, 60 x 30 inches. Courtesy of Henry Urbach Architecture, New York.
24b. Aziz + Cucher, *Chimera #2,* 1998. Color photograph, 60 x 30 inches. Courtesy of Henry Urbach Architecture, New York.

Breeding Better Babies: A New Eugenics?

1. Suzanne Anker, *Material Powers,* 1999. Courtesy of the artist.
2. Charles Kingsley, Illustration for *The Water-Babies: A Fairy Tale for a Land Baby* (1863) (London: JM Dent & Sons, 1957).
3. Frida Kahlo, *Henry Ford Hospital,* 1932. © Banco de México Diego Rivera & Frida Kahlo Museums Trust. From the collection of Museum Dolores Olmeda Patino.
4. Frida Kahlo, *My Grandparents, My Parents and I,* 1936. ©Banco de México Diego Rivera & Frida Kahlo Museums Trust. Courtesy of Los Museos Diego Rivera–Frida Kahlo.
5. Kiki Smith, *Untitled,* 1989. Gampi paper and methyl cellulose, life size. Photograph by Georg Rehsteiner. Reproduced courtesy of the artist and PaceWildenstein, New York, © Kiki Smith.
6. Joel-Peter Witkin, *Portrait of the Holocaust,* 1982. Courtesy of the Fraenkel Gallery, San Francisco.
7. Ariel Ruiz i Altaba, *The Three Graces,* 2002. Courtesy of the artist.
8. Rosamond Purcell, *Stained and Cleared Specimen of Hydrocephalic Child (Tiny Fetus),* 1995 (From the Universiteitmuseum in Utrecht). Photograph © R.W. Purcell 2000. Courtesy of the artist.
9. Marc Quinn, *Lucas,* 2001. Human placenta and umbilical cord, stainless steel, perspex, refrigeration equipment, 80 1/2 x 25 3/16 x 25 3/16 in. (204.5 x 64 x 64 cm). Courtesy of Jay Jopling/White Cube, London, © the artist.
10. Dr. Andrew Goldman, Sonogram of 38-week fetus, now born and named Phineas Goldman. Courtesy of Dr. and Mrs. Andrew Goldman.
11. Lennart Nilsson, Endoscopic photograph, 1965. Courtesy of Lennart Nilsson/Albert Bonniers Forlag AB.
12a. Alexander Tsiaras, 3-D image. Courtesy of Anatomical Travelogue–Alexander Tsiaras.
12b. Alexander Tsiaras, 3-D image. Courtesy of Anatomical Travelogue–Alexander Tsiaras.
13. Adam Fuss, *Invocation,* 1992. Courtesy of the artist and Cheim & Reid, New York.
14. Chris Buck, *Dolly,* 1997. Courtesy of the artist.
15. Laurie Anderson, *What You Mean We?* (scan of video still), 1986. Courtesy of the artist and Canal St. Communications.
16. Beryl Korot and Steve Reich, From *Three Tales* (Act 3 *Dolly*) (video still), 2002. Image of a pipette extracting nucleus of cell. Courtesy of the artists.
17a. Zhang Xiaogang, *Bloodline: Portrait of a Family,* 1993. Courtesy of the artist and Chinese Contemporary, London.

17b. Zhang Xiaogang, *Bloodline: The Big Family, Number 2,* 1995. Courtesy of the artist and Chinese Contemporary, London.
18. Nicolas Rule, *Black Tie Affair,* 1992. Courtesy of the artist.
19. Michael Oatman, *Long Shadows: Henry Perkins and the Eugenics Survey of Vermont (Vermont Pure),* 1995, recreated 2000. Courtesy of the artist and Massachusetts Museum of Contemporary Art. Photo by Arthur Evans.
20. Christine Borland, *A Treasury of Human Inheritance: Huntington's Disease Entres' Case,* 2001. Courtesy of Sean Kelly Gallery, New York.
21. Kiki Smith, *Untitled (sperm),* 1989–1990. Lead crystal, dimensions variable. Reproduced courtesy of the artist and PaceWildenstein, New York, © Kiki Smith.
22. Andreas Serrano, *Frozen Sperm II,* 1990. Courtesy of Paula Cooper Gallery, New York.
23. Iñigo Manglano-Ovalle, *Banks in Pink and Blue,* 1999. Courtesy of Max Protetch Gallery.
24. Hans Danuser, *Tableau I3* from *Frozen Embryo Series I,* 1998–2000. Courtesy of the artist.
25. Faith Wilding, *Embryoworlds* (detail), 1998. Courtesy of the artist.
26. Francesco Clemente, *Silence,* 2001–2002. Courtesy of Robert McKeever/Gagosian Gallery, New York.
27. Catherine Wagner, *Dividing Cell,* 1999. Courtesy of the artist.
28. Heather Sheehan, *egg,* 1999. Photo by Birgitta Homburger. © 2003 Artists Rights Society (ARS), New York. With permission of VG Bild-Kunst, Bonn. Courtesy of the artist.
29. Helen Chadwick, *Nebula,* 1996. © Helen Chadwick Estate, Courtesy of The Zelda Cheatle Gallery.
30. Andy Warhol, *Thirty Are Better Than One,* 1963. © 2003 Andy Warhol Foundation for the Visual Arts/ARS, New York.

Commodification: Genes for Sale

1. Frank Moore, *Oz,* 1999–2000. Oil on canvas over featherboard with roadmap on aluminum frame, 70 x 120 1/2 inches (172,7 x 242,6 cm) overall. From the collection of Howard Stein, New York. Courtesy of Sperone Westwater, New York.
2. Andy Warhol, *Untitled (Tattooed Lady),* 1955. © 2003 Andy Warhol Foundation for the Visual Arts/ARS, New York.
3. Gunther von Hagens, Reclining woman in the 8th month of pregnancy. Courtesy of Prof. Dr. Gunther von Hagens, Institut für Plastination, Heidelberg, Germany. http://www.bodyworlds.com.
4. Gunther von Hagens, Posed whole body specimen with skin. Courtesy of Prof. Dr. Gunther von Hagens, Institut für Plastination, Heidelberg, Germany. http://www.bodyworlds.com.
5. Mark Quinn, *Self,* 1991. Blood, stainless steel, perspex and refrigeration equipment, 81 7/8 x 24 13/16 x 24 13/16 in. (208 x 63 x 63 cm). Courtesy of Jay Jopling/White Cube, London, © the artist.
6. Chrissy Conant, *Chrissy Caviar*®, 2001–2002. (Detail of label). Courtesy of the artist. http://www.chrissycaviar.com.
7. Storey Sisters, *Design 11: Anaphase Dress* (from *Primitive Streak*), 1997. Model: KORINNA @ MODELS ONE. Photographer: JUSTINE. Courtesy of the artists. http://www.helenstoreyfoundation.org.
8. Storey Sisters, *Design 26: DNA Sequence/Spinal Column Dress in Red Silk Chiffon* (from *Primitive Streak*), 1997. Model: KORINNA @ MODELS ONE.

Photographer: JUSTINE. Courtesy of the artists. http://www.helenstoreyfoundation.org.

9. Frank Moore, *Wizard*, 1994. Oil on canvas over wood with pharmaceuticals cast in Lucite in an aluminum frame, 68 x 95 1/2 inches (172.7 x 242.6 cm) overall. From a private collection, Italy. Courtesy of Sperone Westwater, New York.
10. Bryan Crockett, *Ecce Homo*, 2000. Collection of JGS, Inc. Courtesy of Lehmann Maupin Gallery, New York.
11. Bryan Crockett, *Pride, Lust, Gluttony, Sloth, Greed, Envy, Anger*, 2001. Cultured marble. 7 parts, variable dimensions. Edition of 3. Courtesy of the artist and Lehmann Maupin Gallery, New York.
12. Catherine Chalmers, *Rhino*, 2000. Courtesy of the artist.
13. Frank Moore, *Cocktail* 2001. Oil on canvas over featherboard, 15 x 48 inches (38.1 x 121.9 cm). From the collection of Suzanne Slesin and Michael Steinberg, New York. Courtesy of Sperone Westwater, New York.
14. Christy Rupp, *New Labels for Genetically Engineered Food*, 2000. Courtesy of the artist.
15. Adi Nes, *Untitled (The Featherless Chicken)*, 2002. Courtesy of Adi Nes and Dvir Gallery, Israel.
16. Laurel Katz, *The Spotted Merino/Hooked Beetle Sweater Production System*, 1994. Courtesy of the artist.

17a. Larry Miller, *Genetic Code Copyright Certificate,* 1992. Courtesy of the artist.

17b. Larry Miller, *Genomic License.* Courtesy of the artist.

18. Ellen Levy, *Recombinant + DNA*, 2002. Courtesy of the artist.
19. Justine Cooper, *Moist* (video still), 2002. Courtesy of the artist.
20. Catherine Wagner, *–86 Degree Freezers: 12 areas of concern and crisis*, 1995. Courtesy of the artist. Collection of the Museum of Fine Art, Houston.

Science As Culture: Through the Artists' Lens

1. Ronald Jones, *Untitled (Core of the Human Retrovirus),* 1989. (Human T-lymphotic virus Type I which contains protein p25, the RNA which carries the virus's genetic information, and the enzyme reverse transcriptase, corresponding to the viral RNA.) Courtesy of the artist and Metro Pictures.
2. Constantin Brancusi, *The Origin of the World*, 1924. © 2003 Artists Rights Society (ARS), New York/ADAGP, Paris.
3. Frank Gillette, *The Broken Code (for Luria)*, 2002. Courtesy of the artist and Universal Concepts Unlimited, New York.

Deciphering DNA: The Art and Science of a Supermolecule

THE SCIENCE OF GENETICS HAS BECOME A SPECTACLE, a source of multiple metaphors and provocative visual images. In scientific terms, the gene is no more than a biological entity, data without dimension.[1] But in our cultural discourse, this molecule has become a powerful and pervasive icon—an almost magical force—with social meaning that is independent of its biological properties. The language of genetics, the structure of its images, and its social implications are found in news reports, film and television dramas, magazine stories, novelistic plots, and, increasingly, in visual culture. Far more than an arcane technical subject, genetics has become a window on human nature, a cultural narrative, a science with social and figural significance.[2] And its symbols, the chromosome and double helix, have become familiar images, indeed "super images." Like the Mona Lisa, writes art historian Martin Kemp, "They have so insinuated themselves into our public consciousness that they have utterly transcended their original context."[3]

Specific discrete sequences of DNA, called genes, specify the composition of proteins and so carry the information that promotes the formation of living tissues and cells. But, freighted with symbolic and metaphorical associations, genes are perceived as the basis of personal identity, the essence of the self, the determinant of fate. DNA has been called "the ultimate explanation of a human being," the "blueprint of destiny," the "secret of life," a "portrait of who we are." For James Watson, codiscoverer of the structure of this supermolecule, "DNA is the most golden of all molecules."[4] And so in its multiple golden guises, it has achieved celebrity status.

Since the early 1990s, works of art inspired by genetics and biotechnology have become a prominent feature in museums and gallery exhibits in the United States, Europe, and Australia. Universities, art schools, and professional associations have convened exhibitions and symposia on the "sci-art" relationship. Many of these are focusing on the ways artists incorporate genetic concepts into aesthetic strategies.[5] Only obliquely related to the long tradition of medical art, these events are not just about the form of the human body. The artists represented in such venues dwell on the molecule as the code script of life, and they seek to understand the implications of reducing living matter to its genetic essence. They also consider the values that are affected by genetic engineering, new reproductive technologies, and biotechnology applications. They seem to have a special appetite for the issues and associations emerging from recent discoveries in genetics and their ubiquitous public presence.

Artists have always drawn references from their cultural milieu, responding to prevailing issues, social concerns, and cultural narratives.[6] In the past, artists have used science as a source of aesthetic imagery in various ways, ranging from Thomas Eakins's direct depictions of surgical procedures to the reordering of pictorial space by the cubists. In the earliest days of the biological and anatomical sciences, painters and sculptors—for example, Vesalius and Leonardo—sought anatomical exactitude in their delineations of the body, often seeking models in surgical amphitheaters or morgues. Artists and anatomists, writes Deanna Petherbridge, "met around the dissecting table," and their precise drawings made a significant contribution to the growing scientific understanding of the human body and its structure.[7] For example, the 600 richly illustrated pages of Andreas Vesalius's 1543 anatomy treatise, *De Humani Corporis Fabrica Libri Septem*, played an important role in the rudimentary science of anatomy. And anatomists collaborated with artists to produce often fanciful images of the body.[8] Indeed, at that time it was difficult to define the boundary between science and art.

Technological developments in optical instrumentation enabled artists as well as scientists to extend their ways of seeing and their visual descriptions of the body. But new visualizing techniques also expanded the possibilities for imagination and play. The invention of photography in 1839 had a profound effect on recording our ways of seeing. Not only could an image be captured in a permanent form, but it could be widely circulated. Also, the increased use of the microscope and later of X-rays and other imaging technologies opened an invisible world to the imagination of artists, a world beyond the naked eye.[9] In his novel *The Magic Mountain*, Thomas Mann played on the importance of these instruments when Hans Castorp swoons over the chest X-rays of the tubercular heroine, Claudia Chauchat.

Evidence of this "quest for invisible reality" also became an underlying idiom in the work of visual artists such as Kasemir Malevich, Frantisek Kupka, and Marcel Duchamp.[10] Then, as new computerized imaging technologies—MRIs, CAT scans, and PET scans—further enhanced the visual arena of scientists, these images entered the domain of the visual arts. The fragmented and telescopic views generated by instruments of science were reflected in the "cut and paste" techniques of collage and montage. From the microscopic, to the macroscopic, artists deconstructed the familiar in their pictorial schemes. And, reflecting the more complex understanding of "reality," artists moved toward abstraction.[11]

Why the gravitation toward genetics—a complex and highly abstract science, a science of the invisible—with little apparent visual appeal? For one thing, today's artists are perpetuating the established aesthetic tradition of curiously engaging scientific concepts and iconography to convey their perspective on the changing world around them. Artistic interpretations in the past have provided insight into the social impacts of the most critical sciences and technologies of their day. In the early 20th century, the splitting of the atom influenced the work of early European abstractionists such as Wassily Kandinsky and Piet Mondrian.[12] In the 1940s, American abstract painters such as Mark Rothko and Barnett Newman expressed their vehement reaction to the atomic bomb by creating biomorphic images of a primordial world.[13] Pop artists Andy Warhol and Robert Rauschenberg commented on technological progress by incorporating inventions such as the electric chair and the X-ray into their art. Today, artists are responding to the deluge of discoveries in genetics by integrating the images evoked by this science into their concepts and styles of figuration.

Genetics is surely one of the most critical sciences at the turn of the 21st century; its exploration of inner space has competed for media attention with the exploration of outer space. In public communications, scientists themselves suggest they are revealing the canonical definition of what and who we are, and they often endow the genes with social as well as physical attributes. Purported obesity genes, gay genes, risk genes, violence genes, genes for directional ability, genes for saving, and even genes for sinning are then trumpeted in scientific venues and in the media.[14] Repeated media messages have amplified the metaphorical meanings of genetics. Some reports suggest that behavioral characteristics and personal identity are encoded in a molecular text, that the most complex human attributes are caused by genetic predispositions, and that geneticists will soon acquire the awesome power to manipulate DNA and thus to determine humanity's fate. These social as well as scientific beliefs and expectations have shaped the cultural context in which the "molecular gaze" has emerged in the world of contemporary art.

The interest in genetics among contemporary artists reflects a preoccupation with the human body, the corporeal self, and the essence of human nature. In our present epoch, technologies such as genetic testing, genetic engineering, cloning, and "reprotech" all involve the transformation of the body. The growing possibilities of altering the body, tampering with nature, and manipulating reproductive processes are clinically and philosophically seductive, yet troublesome as well. They promise control and even perfection, but they also evoke fundamental questions of authenticity, identity, and bodily integrity—the same questions that, two centuries ago, inspired Mary Shelley to create Frankenstein.[15]

Both scientists and artists are exploring the nature of personhood, the meanings embedded in the corporeal body. Both are probing the deeper world underlying surface manifestations. Both aspire to comprehend nature's secrets. By redefining the body and reducing it to an array of molecules and their sequences, geneticists are changing our understanding of human nature. The "living text" in this paradigm becomes a system, composed of assemblies of molecular parts responding to recipes, instructions, and codes.

Challenging the conventional and widely assumed boundaries between life and death, human and machine, the self and the other, the molecular model implies a radical shift in our visions of what it means to be human. Although the understanding brought about by this genetic revolution opens possibilities for medical and forensic benefits, it has also created social uncertainties, ethical dilemmas, and personal anxieties—and these too are reflected in visual art.

The gaze of artists often dwells on the values challenged by a rapidly growing science that frequently seems to defy natural categories, common morality, and traditional understandings of human nature. Touching on the moral and ethical dilemmas of manipulating nature, the patenting of genes, and the concerns brought about by the changing status of humans in the post-genomic world, visual artists portray the expectations and salient anxieties of our genetic age.

Such depictions of science matter. They are important in bridging the distinction between the work of science and the world of cultural meaning. They express the concerns of the public and shape societal responses. More accessible than scientific publications, visual images turn the technical talk of science into the emotional domain of public discourse. Thus, they help to arrange the cultural assimilation of the genetic revolution. Through its creativity and awareness, writes Valerio Deho, curator of the biennial exhibition DNArt, "Art can render technology intelligible to the public and allows a process of self-reflection to begin."[16]

As artists interrogate science, they can also form political conceptions. Works of art, writes sociologist Murray Edelman, provide us with models, scenarios, and images that help us to make sense of the world.[17] Whereas historians often focus their analysis on the influence of social conditions on art, the images of art, writes Edelman, also help to construct social ideas and transform public beliefs.

Geneticists, in fact, are increasingly aware of the importance of art. They frequently exhibit artists' images of molecular structures on walls of their institutes and the covers of their journals.[18] Science journals such as *Nature* and *The Lancet* publish regular articles and commentaries on visual art. Martin Kemp writes a column in *Nature* about the relationship of science and art.[19] Some scientists are turning their own scientific images, derived from their instruments, into works of art. The Wellcome Trust, primarily a science-funding organization, has organized a "sci-art" Program to encourage collaborations between artists and scientists.[20]

Artists' visualizations not only expand the audience of science, they also challenge familiar and unquestioned assumptions by extending beyond the constraints of the present to imagine alternative possibilities.[21] A major scientific endeavor like the Human Genome Project is hegemonic. As a dominant cultural narrative, it has institutionalized a particular concept of reality that appears to be exclusive and inevitable. It has influenced health care, jobs, and legal institutions. Increasingly ubiquitous, genetic explanations of health, human behavior, and personality pervade both the medical and social discourse, while tending to exclude other possible interpretations and explanations. "It's all in the genes."

Art can convey other conceptions of reality, providing countervailing visions and calling attention to ambiguities that pervade everyday life. Writes Edelman, "The very production and existence of serious art encourages skepticism towards the pieties of regimes...raising doubts that orthodox interpretations of the social world should be regarded as the only valid ones."[22] Art, often shocking and distorting, can de-familiarize the obvious and so disturb the canonical peace.[23]

As we show in the pages that follow, the images evoked by the genetic revolution build on certain archetypal forms (such as monsters and chimeras), fundamental themes (such as portraiture and identity), and common techniques (such as montage and collage) that have appeared and reappeared throughout the history of art. In this book, we explore how artists have reworked these traditional genres in response to genetic ideas and information. How has this science been integrated into the visual culture? How do artists deal with the fantasies that are expressed by geneticists themselves as they promote the importance of their often speculative science and envision its future applications? How do artists portray the diverse social, symbolic, and ethical implications of the genetic revolution?

These questions provide a frame for *The Molecular Gaze.* We have arranged the images according to themes that we have found to be prevalent and relevant. We have chosen not only images that illustrate and represent genetics and the broader ethos of the genetic revolution, but also images that hold their own in the world of art. As a collaboration between a visual artist and a sociologist of science, our text is intended to place these visual images in their social context and to provide selected background information about the science, its applications, and the debates swirling around its implications. Many of the museum and gallery exhibitions taking place over the past decade present powerful and diverse images.[24] We want to suggest that these images also provide insight into the complex relationship between art and science and, more broadly, between art and its social context. *The Molecular Gaze* organizes and analyzes this art systematically in terms of five predominant themes we see reflected in this new genre:

- The reduction of the body to a molecular text, to a language, a "code script" of information.
- The meanings of mutation, manipulation, and monsters: portraying the new grotesque in contemporary art.
- The blurring of boundaries: transgenics and the re-emergence of the chimera as an image in art.

- The potential for a new eugenics as techniques of assisted reproduction and cloning converge with the possibilities of genetic manipulation.
- The commodification of the body: patenting and the concept of cells and genes for sale.

As an interdisciplinary work, *The Molecular Gaze* has three objectives:

- To explore an increasingly important movement in contemporary art by documenting the range of ideas and images that artists are associating with the genetic revolution.
- To suggest, through depictions of historical motifs, the ways in which DNA representations relate to archetypal images and themes that have appeared throughout the history of art.
- To provide insight into the social and cultural meaning of a consequential and controversial science through the gaze of artists as they interpret scientific processes and engage in cultural critique.

Our task in the following pages is to draw connections among these diverse responses to the genetic revolution in contemporary art.

Notes

1. DNA (deoxyribonucleic acid) is essentially a string of chemical units called bases that embody the genetic code. These chemical building blocks are adenine, cytosine, guanine, and thymine, hence the common acronym ACGT. They occur in slightly varied combinations in different people.
2. Dorothy Nelkin and M. Susan Lindee, *The DNA mystique: The gene as cultural icon.* W.H. Freeman Press, New York, 1995. See also Jose van Dijck. *Imagenation: Popular images of genetics,* New York University Press, New York, 1998.
3. Martin Kemp. 2003. The Mona Lisa of modern science. *Nature* 421: 416–420.
4. James Watson, *The double helix.* Atheneum, New York, 1968.
5. See Appendix.
6. Arnold Hauser. *The sociology of art,* University of Chicago Press, Illinois, 1982; Vera L. Zolberg. *Constructing a sociology of the arts.* Cambridge University Press, Massachusetts, 1990. Arthur Danto. *Encounters and reflections: Art in the historical present.* Farrar, Straus, and Giroux, New York, 1990.
7. Deanna Petherbridge and Ludmilla Jordanova, *The quick and the dead: Artists and anatomy.* University of California Press, Berkeley, 1997.
8. Michael Sappol, "Dream Anatomy." Exhibition at the National Library of Medicine, Bethesda, MD, November 2002–June 2003. Winter 2003. At: www.nlm.nih.gov/exhibition/dreamanatomy/. See also Brian Vastag. 2002. Anatomy exhibit shows charm of the grotesque, *J. Am. Med. Assoc.* 288: 2525–2529 and Deanna Leamon, "Remembering our mortality: Artists and doctors in the gross anatomy lab," paper delivered at the 16th Annual National Conference on Liberal Arts and the Education of Artists, School of Visual Arts, New York, October 2002.
9. Betty Anne Kevles. *Naked to the bone: Medical imaging in the twentieth century.* Rutgers University Press, New Jersey, 1997.

10. Linda Dalrymple Henderson. 1988. X Rays and the quest for invisible reality in the art of Kupka, Duchamp and the Cubists. *Art Journal* (Winter) 47: 323–340.
11. Lynn Gamwell. *Exploring the invisible: Art, science and the spiritual.* Princeton University Press, New Jersey, 2002.
12. John Golding. *Paths to the absolute: Mondrian, Malevitch, Kandinsky, Pollock, Newman, Rothko and Still.* Princeton University Press, New Jersey, 2001.
13. Jeffrey Weiss. 1983. Science and primitivism: A fearful symmetry in the early New York school. *Arts Magazine* (March) 57: 81–87.
14. Dorothy Nelkin. 2001. Molecular metaphors: The gene in popular discourse. *Nat. Rev. Gen.* 2: 555–559.
15. Jon Turney. *Frankenstein's footsteps: Science, genetics, and popular culture.* Yale University Press, New Haven, 1998.
16. Announcement of Biennale Merano arte-DNArt, curated by Valerio Deho, June 15, 2002.
17. Murray Edelman. *From art to politics: How artistic creations shape political conceptions.* University of Chicago Press, Illinois, 1995.
18. For example, the Cold Spring Harbor Laboratory displayed the work of artist Kevin Clarke; the New Academy of Sciences regularly curates exhibits about art and science; *Nature Genetics Reviews* features molecular paintings on many of their covers (see www.nature.com/reviews.)
19. His columns have been collected in Martin Kemp, *Visualizations: The Nature book of art and science.* Oxford University Press, New York. 2000.
20. Edelman, op. cit.
21. Anthony Amsterdam and Jerome Bruner. *Minding the law.* Harvard University Press, Cambridge, 2000, pp. 17 and 235.
22. Edelman, op. cit., p. 67.
23. Amsterdam and Bruner, op. cit., p. 237.
24. Boxer, Sarah. 2003. *At play with DNA. The New York Times.* March 14, weekend edition, p. 1.

Reductionism: The Body as a "Code Script" of Information

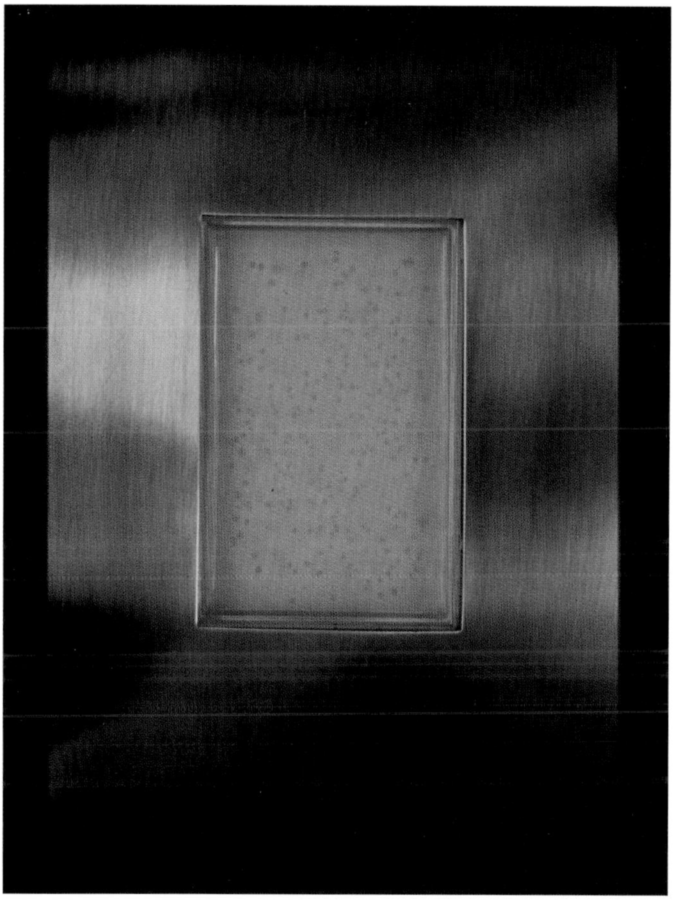

1 Marc Quinn, *Sir John Sulston: A Genomic Portrait, 2001.*

In September 2001, the National Portrait Gallery in London unveiled a highly abstract and unusually strange portrait created by conceptualist, Marc Quinn. This work, *Sir John Sulston: A Genomic Portrait* (Fig. 1), depicts the former director of the Wellcome Trust Sanger Center in Cambridge UK.[1] Sulston had played a critical role in the international effort to map, then sequence, the human genome, leading the UK's effort in human genetics research. Quinn's portrait of Sulston is not in the least way traditional. It comprises colonies grown from bacterial cells taken from Sulston's sperm that contained segments of his DNA. Placed in a flawlessly finished mirror-like frame, this portrait reflects the gazing viewer's image while encasing a centered overall field of creamy blots. Patterns of migrating cells seem to appear magically.

To Quinn, this portrait of Sulston is completely realistic, "the most realistic portrait in the Portrait Gallery," because it carries the actual genetic instructions that led to the embodiment of the subject. The artist describes the piece as "a portrait of his parents and every ancestor he ever had back to the beginning of life in the universe. I like that it makes the invisible visible and brings the inside out." However, to the subject, Sir John Sulston, "The portrait contains a small fraction of my DNA, so it's only a detail of the whole, though there is ample information to identify me."[2]

Are people merely the measure of their genes or simply products of history—reduced to their personal experience, social relationships, private memories, and specific cultural values? The segments of DNA comprising Sulston's portrait do indeed reflect his biological history, but what exactly creates his unique identity? His personhood? Is the self simply a sum of its biological parts, or is it a dynamic, fluid interactive system shaped by culture and mutable over time? Biologists in the molecular age are seeking to answer the degree to which these questions address the crucial characteristics of life. Exploring life at its molecular level has often enforced causal claims of genetic determinism.[3]

Responding to, while endorsing, mapping the human genome, artists like Marc Quinn invoke molecular images to query the meanings of human identity. Other artists similarly question the significance of reductionist premises governing contemporary biological sciences.[4] They appropriate signs and symbols of genetics, the chromosome and double helix, to visually explore the linkages between the supermolecule DNA and corporeality, pathology, and social transformation.

Reading the Body

The quest to understand what makes us human is hardly an original endeavor in the worlds either of science or of art. Nor are reductionist explanations. The 19th-century sciences of phrenology and physiognomy postulated that facial and bodily shapes were physical indicants of personality and mental states, and that they could be used as clues to understanding human behavior.[5] The body was transformed, became a surface, in effect a text, on which physicians and scientists inscribed and projected cultural meanings that had powerful implications for social policies.[6]

In 1807, the German anatomist Franz Joseph Gall (1758–1828) published the foundational text in phrenology, visually mapping behavioral characteristics and psychological attributes to the geography and physiology of the brain. Bumps and depressions on the surface of a skull, he believed, were indicators of inherent behavioral characteristics. His phrenology diagram (Fig. 2) causally connects specific regions of the brain to specify talents

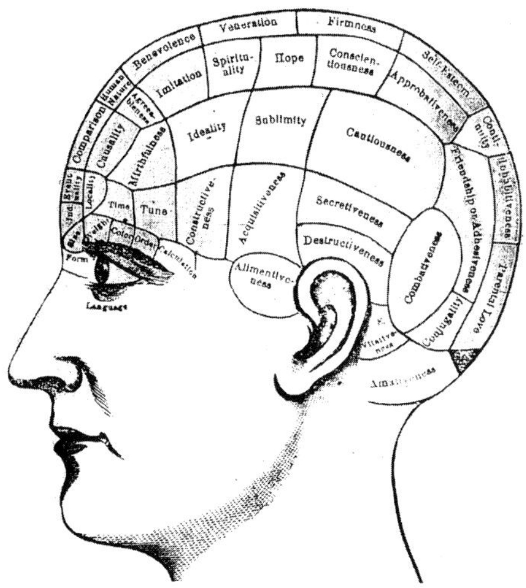

2 Franz Joseph Gall, Phrenology Diagram.

and dispositions. Although his concept of brain localization held some validity, Gall's analysis of function was fatally flawed. For example, he assigned a number to each area of the brain corresponding to its hierarchical importance in the cerebellum. Interestingly, Gall assigned the number-one position to sex—an idea preceding and possibly influencing the later psychosexuality theories of Sigmund Freud.[7] Although by the 1850s phrenology had been rejected as a legitimate science, it continued to influence popular understanding about the legibility of the body.[8]

In the latter part of the 19th century, Italian criminologist Cesare Lombroso applied these theories to criminal behavior. Well into the 20th century, criminal investigators and forensic scientists relied extensively on visualizations of facial and body shapes to read inner truths about character, personality, and fate. Photographs and drawings were commonly employed as a way to identify criminal types, maniacs, or garden-variety mentally ill.[9]

This idea that the body could be read as a character map had precedence in the early science of physiognomy and in its representation in Charles le Brun's 17th-century drawings.[10] le Brun, first painter to King Louis XIV, developed a model of "reading" man's inner being by comparing human and animal faces, placing them side by side in detailed drawings. For example, in *Visage* (Fig. 3), a bull's emotions are linked to the physical features of man's bodily characteristics. In this case, the wide-eyed human creature has extended ears, elongated muzzle, and diminished forehead, the very physical characteristics anatomically present in a bull. le Brun endows the man with an expression bordering on the primeval, an instinctual brute, but a sentient being. In other comparisons between men and animals, the artist associates

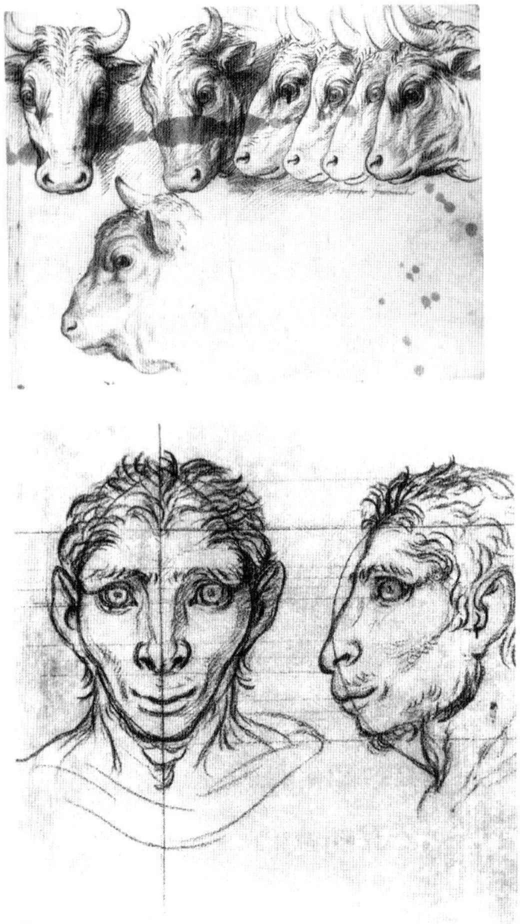

3 Charles le Brun, 17th century. *Top: Etude de Taureau et de Boeuf. Bottom: Visage.*

the physical characteristics of cats with stubbornness and mistrustfulness, and curved noses are seen to be indicative of eagle-like men, having the aggressive character of birds of prey.[11] The parallels suggest that passions and personality types can be decoded through facial features.

As the sciences of phrenology and physiognomy became assimilated into 19th-century popular culture, artists were increasingly drawn to the analogy between inner character and external facial and bodily characteristics. Daumier's drawings and bronze busts from the 1830s also build on the prevailing assumption that the body is a social text.

Daumier portrayed witty caricatures of various devious and pretentious social types, from the politician (Fig. 4a) to the mockingly posed janitor (Fig. 4b).[12] He exaggerated the sizes of noses, foreheads, and postures to satirize social class stereotypes. Their specific body parts are looked upon as indicators of evolutionary origins.

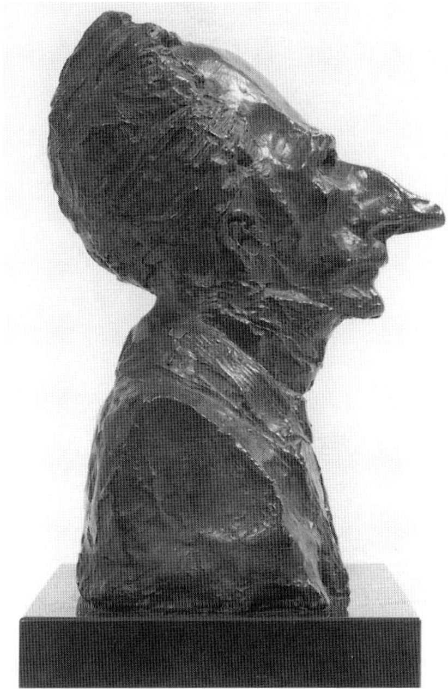

4a Honoré-Victorin Daumier, *Jacques Lefebvre: The Fine, Trenchant Spirit* (*L'Esprit fin et Tranchant*). ca. 1832–1835.

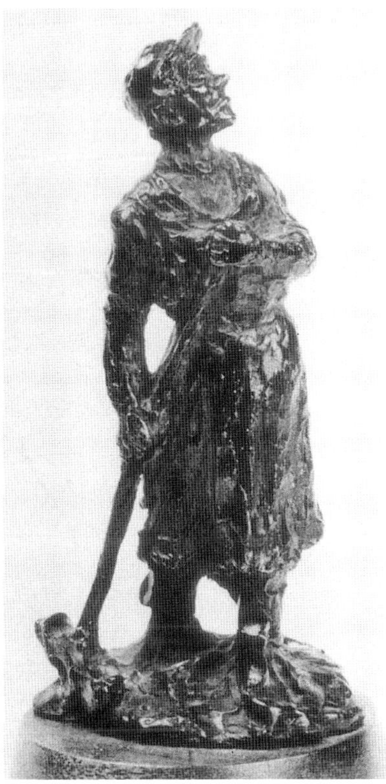

4b Honoré-Victorin Daumier, *The Parisian Janitor* (*Le Portier Parisien*), ca. 1840–1862.

Similarly, Degas was profoundly influenced by Lombroso and the "science" of physiognomy. In his portraits of ballet dancers, he explored the physical attributes of working-class girls. *The Little Fourteen-year-old Dancer* (left profile) (ca. 1880, cast 1922) (Fig. 5), according to 19th-century art critic Paul Mantz, disturbed contemporary aesthetes because the contours of the dancer's face exposed her "animal impertinence." Her protruding lower jaw and receding forehead were shapes widely read as signs of primitive traits with atavistic associations.[13] In case there were doubts about Degas' intentions, he exhibited the sculpture alongside his pastel sketches of criminal heads.[14]

Nineteenth-century artists were also influenced by the growing popular interest in Social Darwinism—the belief that socially deleterious traits were directly inherited. Providing biological and hereditarian explanations of a very wide range of social behavior, Social Darwinist beliefs laid the conceptual foundation for the theories of degeneration underlying the eugenics movement.[15]

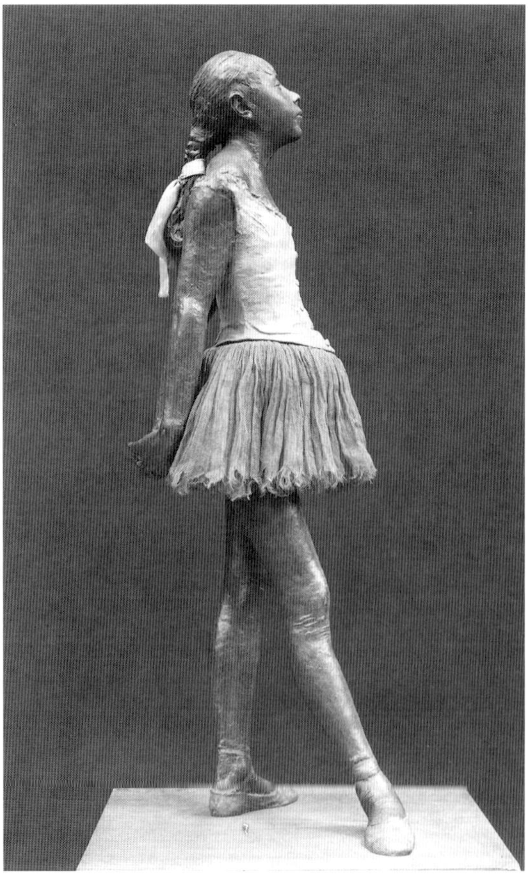

5 Edgar Degas, *The Little Fourteen-year-old Dancer* (left profile), ca. 1880, cast 1922.

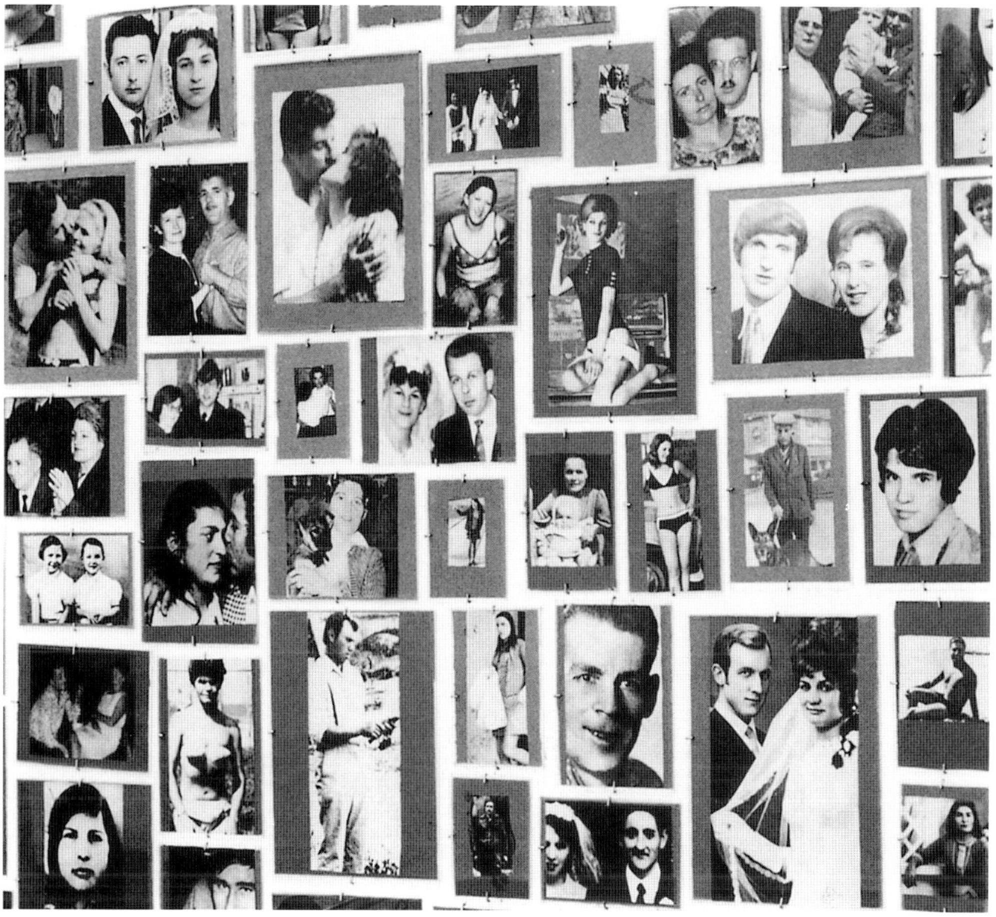

6 Christian Boltanski, *The Detective* (detail), 1972–1973.

During the eugenics movement, photographic images of the body played a dominant and documentary role. Eugenicists portrayed "degenerate" bodies with biological abnormalities, encoding them, writes historian Nicole Hahn Rafter, with visible signs of evil: "The ape-like born criminal [was] a monstrosity who manifested wickedness—from coarse head hair to prehensile toe."[16] Photographs of the faces and bodies of "degenerates" were frequently deployed as visual narratives to support eugenic ideas and justify social practices that would elevate "the innate capacities of [the] citizenry."[17] Nearly a hundred years later, French artist Christian Boltanski revisited the assumptions underlying this reading of the body by pairing photographs of murderers and their victims. He asks viewers to differentiate who was who. But both appear so genial and so human that one could not possibly discern their character alone from looking at their portraits (Fig. 6).

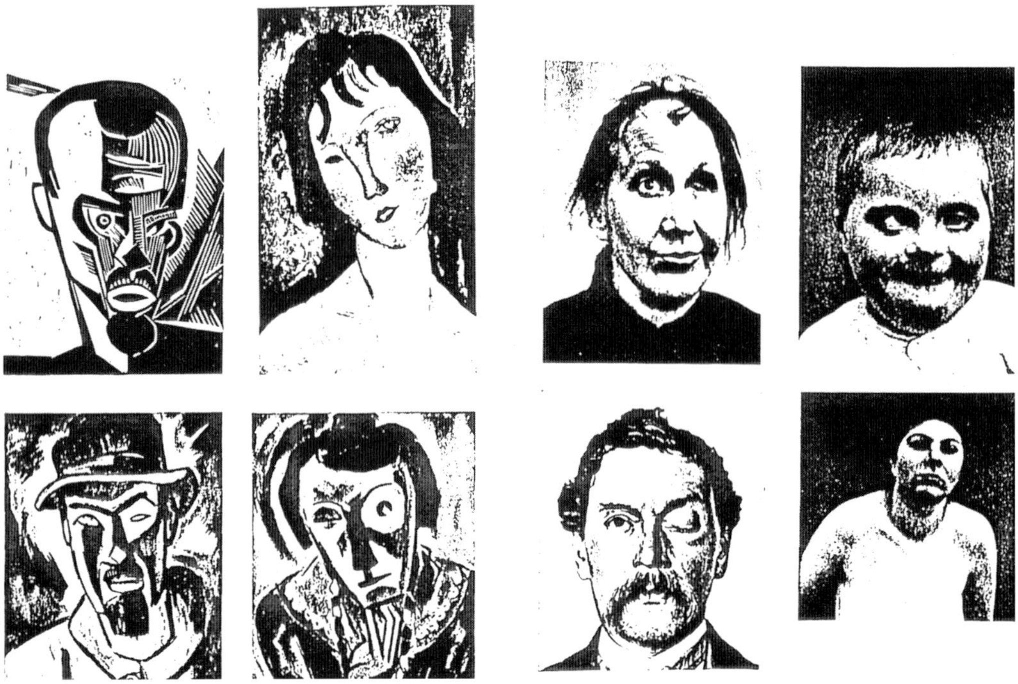

7 Juxtaposition of works of "degenerate" art by Karl Schmidt-Rottluff and Amedeo Modigliani and photographs of facial deformities, from Paul Schultze-Naumburg, *Kunst und Rasse*, 1928.

Convictions about the meanings to be read in the body persisted in Germany during the 1930s. The National Socialist Government of Germany associated modern art with degeneration, defining it as a pathological reflection of a dysgenic society. Paul Schultze-Naumburg, in his 1928 treatise *Kunst und Rasse* (*Art and Race*), drew analogies between modern art and physical deformity by juxtaposing works that were considered "degenerate" with photographs of physically diseased people (Fig. 7). The visual analogy was striking, yet utterly misleading. Whereas the pictures on both panels portray deformity and exaggeration of facial features, the artists, Karl Schmidt-Rottluff and Amedeo Modigliani, whose works are represented in the left-hand panel, had consciously created the distortions on their subjects' faces. Schmidt-Rottluff was reflecting the conventions of avant-garde artists in the German Expressionist movement. Modigliani's portraits employed elongation, a type of distortion associated with tribal cultures. For these artists, expression rather than verisimilitude was of prime importance, and their twisted faces were ways to portray both an internal, subjective world and the spirit of the times. In contrast, the facial deformities represented in the right-hand panel were "acts of nature" that could not be willfully changed. Precursors of the later ideology of the Fascist regime, writers such as Naumberg and Max Nordau, linked both physical deformity and distorted representation to a sick society.[18]

Joseph Goebbels' Ministry for Popular Enlightenment and Propaganda during the Third Reich developed state policies based on this analogy between modern art and degeneration. The ministry attacked the very notion of modern art and promoted works that were believed to represent the authentic nature of the German people. In Munich in 1937, the National Socialist regime mounted two exhibitions. *Entartet Kunst*, otherwise known as the Degenerate Art show, consisted of approximately 650 works of art that were labeled "un-German" and, therefore, racially inferior.[19] The purpose of this show, which included paintings by Braque, Ernst, Klee, and Picasso, and even many German artists such as Nolde, Marc, Beckmann, and Kirchner, was to humiliate modern artists by showing their inability to represent the external world accurately.

The second exhibition, simultaneously held in Munich's newly constructed Haus der Deutschen Kunst, was entitled *Grosse Deutsche Kunstausstellung*. This show consisted of paintings and sculpture in the idealized image of allegedly classical and purified beauty. This official art of the Third Reich borrowed images from Greco-Roman statuary and neoclassical painting. It was intended to portray German society as the essence of what Hitler and his Nazi ideologues referred to as the pure Aryan ideal.[20]

The Language of the Molecular Vision: Reading the Genetic Code

In his seminal 1943 book, *What is Life?*, the physicist Erwin Schroedinger identified a rhetorical model of the gene, calling it a "code script," and referring to chromosomal structures as "law-code and executive power."[21] Such language implies that an organism itself, through its machinery of building blocks—cells—can be decoded, stripped of nebulous mystery.

The idea took off in the early 1950s, when the mathematical precursors of information theory and electronic communications technologies began to permeate and reshape the biological sciences.[22] Questions about the definition of life provoked the interest of physicists who sought to extend their discipline toward a new frontier.[23] Applying the laws of matter to living systems, for example, George Gamow recognized that biological information could be viewed as information storage and retrieval systems and thus presaged the ability to store DNA sequences in searchable databases. The impact of digital control systems on representations of life processes would radically transform the biological sciences in ways that were later to influence a direction of the visual arts.

The language of information theory, developed in the 1950s by physicists to deal statistically with the content of information, implied that living things were assembled according to instructions encoded in the chromosomes. Developing an information-based biology, geneticists worked on the assumption that organisms could best be understood by decoding these instructions. The images of pathology, the marks of human difference, thus moved from gross to hidden body systems, from facial and bodily shapes to genes. This re-conceptualization of the body, from a morphological structure to a molecular organization, in effect converted flesh and blood into information. Perceiving the body as a decipherable text, scientists increasingly employed linguistic tropes and communication models in their descriptions.[24] For example, when Francis Crick and James Watson revealed the double-helix structure of DNA in 1953, they described the gene sequence as a "master molecule," an "information system," a "code which carries the genetic information."

At the molecular level, the four nucleotide bases comprising DNA are signified by four alphabet letters—ACGT—in infinitely varied recombinations. These letters, representing Adenine, Cytosine, Guanine, and Thymine, code for units of biological systems from simple bacteria to complex plants and animals. Living systems are considered a set of "instructions," a "program," or a "master code" transmitted from one generation to the next.

Language metaphors are fundamental to the discourse of the biological sciences.[25] In 1967, Robert Sinsheimer authored *The Book of Life,* in which he likened the genetic code to the ancient Mayan Codex.[26] Biologist Erwin Chargaff described chromosomes as "the grammar of the body," forming all the instructions for our genetic makeup.[27] Extending the metaphor, biologist Robert Pollack, in his 1994 book, *Signs of Life,* discussed the cell machinery as a sort of word processor. "To understand the linguistic properties, syntax, grammar, and semantics of a human genome, we have to look at how a gene speaks to a cell and at the meanings of what it says." But Pollack also insisted on the importance of subjective interpretation. "Genomes with their networks of interactions and their multiplicity of meanings leave us free to use our imagination as we read them."[28]

Linguistic metaphors imply that the cell has a vocabulary of code words with which it constructs and receives messages.[29] They are useful in exchanges among scientists who use such words as "transcription" and "translation" to describe the actions of DNA and RNA in the cell. They are also useful in communicating this complex science to the public and in promoting the value of genetic explanations. Biologist Bruce Wallace likens the human genome to "the torn pages of a giant novel, written in an unknown language, blowing about helter skelter in an air conditioned enclosed space such as Houston's astrodome."[30] Scientists involved in mapping and sequencing the genome, so this metaphor implies, will capture the pages, put them in proper order, and therefore be able to analyze the meaning of the resulting text. In a controversial interpretation, living organisms are simply reduced to "readouts" of their DNA. DNA, claims James Watson, is "what makes us human."[31]

Molecular vision has increasingly dominated the assumptions and methods of the biological sciences. Reducing life itself to molecules, it has displaced the visceral references that had once defined the authenticity of the body and the authority of traditional biology as a descriptive science. Despite the complexity of life, this vision implies that we are but a sequence of nucleic acids, a "code script" of information.

This transformation of biology from organism to code and/or text parallels developments in art. Artists are adapting images revealed through high-technology apparatus, and their pictorial and sculptural products have shifted toward the abstract. They have recognized in genetic iconography an underlying narrative that resonates with familiar forms and issues in the history of art.

The Integration of Language into Modern Art: Historical Considerations

Ancient scripts—from inscriptions on oracle bones, to cuneiform tablets, to the Rosetta Stone—offer insight into the origins and evolution of language. They document a transformation from the pictogram, to the ideogram, to the phonetic alphabet. Operating as a communication system, writing originated as pictorial representations, but ultimately broke away from descriptive depictions, evolving into symbol sets with conventional and universally understood meanings.[32]

Interested in archaic and lexical forms as visual clues to the past, Paul Gauguin, in *Tehamana Has Many Ancestors* (1893) (Fig. 8), reflects his passion for remote places, such as the famously exotic island of Tahiti. In this painting, the viewer is presented with a past still visible in the present.[33] Gauguin juxtaposes Rongorongo writing, that is, pictograms from Easter Island, with a portrait of his young mistress dressed in western-style clothing. She belongs, in effect, to two distinct cultures. As evident in the title of the work, *Tehamana* evokes the importance of her genetic ancestry. However, the field of pictograms in the background also draws distinct attention to the artifacts of cultural evolution, invoking the significance of heredity in a distinct context of cultural change.

Exploring ways to uncover and depict the inner self, surrealists during the 1920s experimented with automatic writing or *ecriture automatique*, a form of expression over which there is no critical or conscious control. In his 1924 Manifesto of Surrealism, Andre Breton declared this artistic method as "pure psychic automatism,"[34] a way to visually expose the unconscious. As an example, Henri Michaux's early painting, *Alphabet* (1927) (Fig. 9), consists of purely invented signs laid out in chart-like rows. Undecipherable as a workable alphabet, these linear musings share their rhythm with writing as

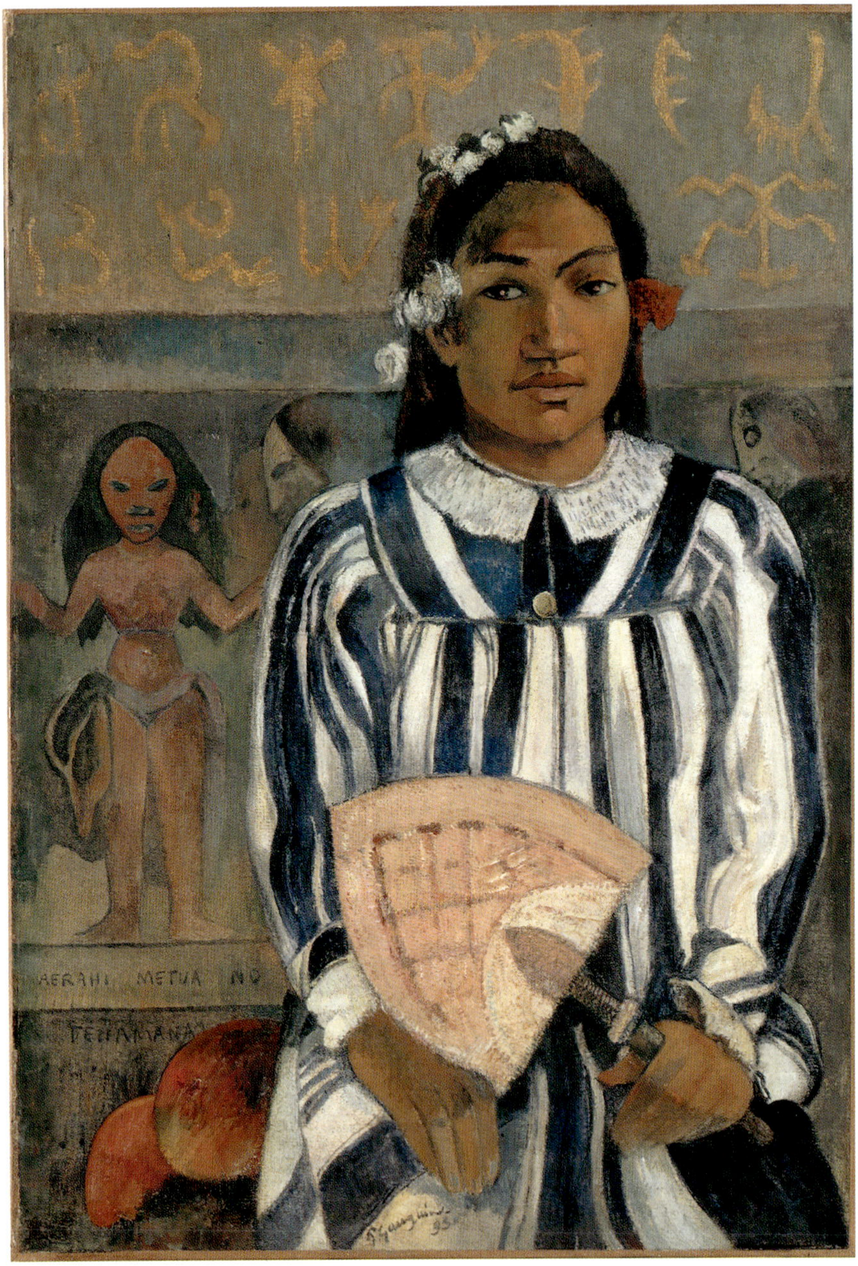

8 Paul Gauguin, *Tehamana Has Many Ancestors, (Merahi metua no Tehamana)*, 1893.

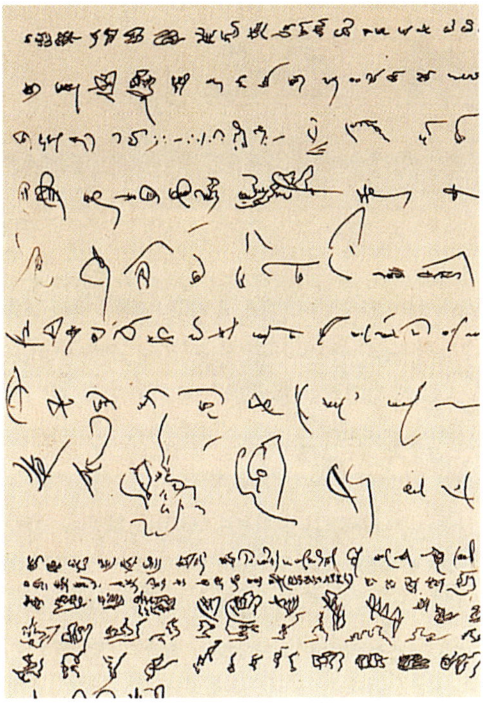

9 Henri Michaux, *Alphabet* (detail), 1927.

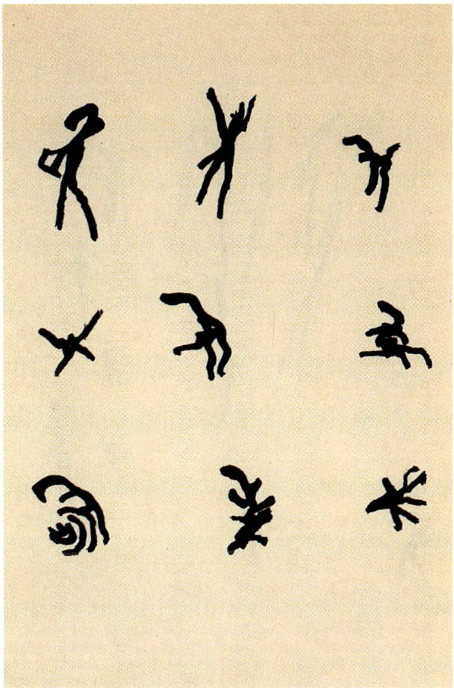

10 Henri Michaux, *Untitled (Movements [Mouvements])*, 1950–1951.

concrete expression of feeling and thought. This poetic choreography consists of a series of elegantly syncopated graphic symbols. Later, in the 1950s, Michaux embarked on a series of approximately 1200 ink drawings called *Untitled (Movements [Mouvements])* (Fig. 10), which the artist described as "interior gestures." Michaux's figural ideograms, in their calligraphic arabesques and animated scale, resemble chromosomes, language, and body forms. He drew them in "a state of innocence," describing them as automatically emerging from his unconscious, free from the constraints of words.[35]

The unconscious process of image-making developed by the surrealists was later to influence the abstract expressionists, particularly Jackson Pollock, who also integrated language signs in his paintings. In *Stenographic Figure* (1942) (Fig. 11), the artist, with graffiti-like exuberance, calligraphically inscribes abbreviated phonetic signs atop a reclining body. These gestures can be read as an X, a 6, a zero, all of which are recognizable in Pitman stenography, or shorthand (Fig. 12).[36] In shorthand, these discrete mark-

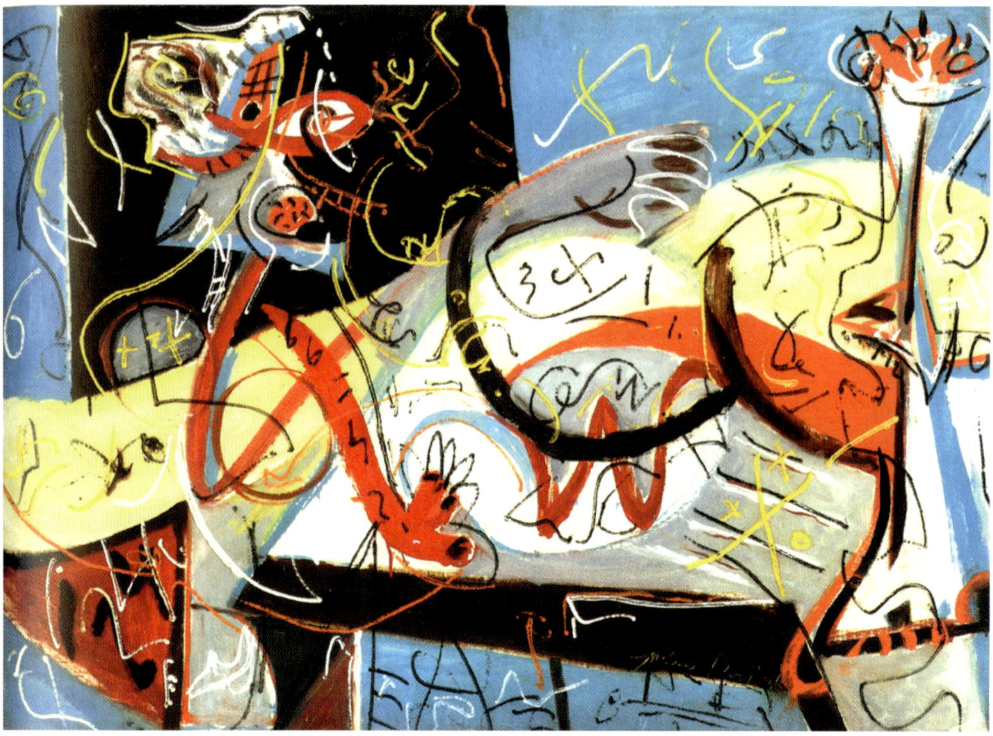

11 Jackson Pollock, *Stenographic Figure*, 1942.

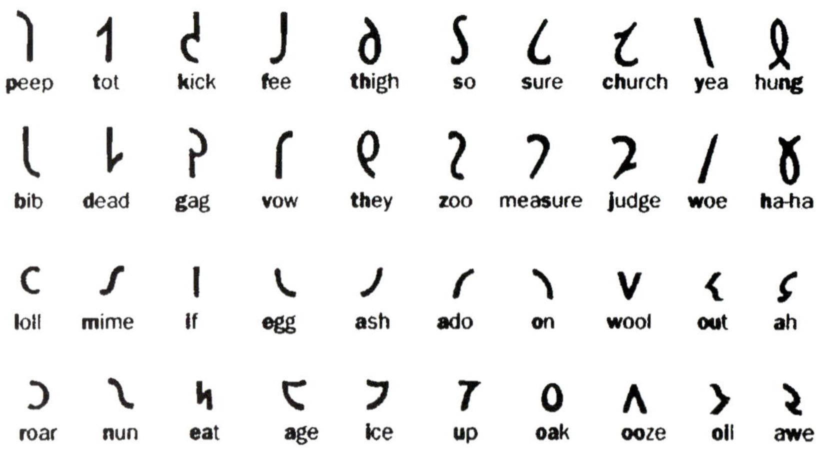

12 Tracy Wellman, *Kingsley Read's Shaw Alphabet* (after Pitman).

ings, known as phonemes, are the smallest units possible in any system of language. Employed as a rapid means of communication, this writing system substitutes symbols for letters. Pollock thus transforms a bodily entity into abstraction through the use of this reduced language. By way of analogy, the symbols that he uses can be compared to the ways in which the smallest units of hereditary material, the genes, form the text of the body.

Salvador Dali was the first artist to explicitly represent the double helix in his work. Only a few years after Watson and Crick discovered the structure of DNA, he created several paintings using this image. In *Butterfly Landscape, The Great Masturbator in Surrealist Landscape with DNA* (1957–1958) (Fig. 13) and *Galacidalacidesoxyribonucleicacid* (1963), Dali's double spiral manifested "the only structure linking man to God."[37]

For Yves Klein, the body itself became a painting tool, a mark-making device. In his *Anthropometry of the Blue Period* (Fig. 14), a 1960 performance piece held at the Galerie Internationale d'Art Contemporain in Paris, Klein had three nude models cover their bodies in paint. The models, whom he referred to as "living paintbrushes," then pressed their figures against paper laid out on the floor.[38] As a direct transfer, the body prints are in schematic form, but they nonetheless read as an external physique. Scaled down, they appear as "colored bodies," an etymological definition of chromosomes themselves.

Chromosome-like forms also characterize the bark paintings (Fig. 15) made by the Mbuti, one of the last remaining hunter-gatherer tribes in the Congo, the former Zaire. With this array of scattered signs, spontaneously and asymmetrically placed, the Mbuti artists employ decoration as a way to order and classify their world. Bark cloth becomes visually alive, a cacophony of symbols representing their cultural beliefs.[39] Ranging from a stick figure design to X- and U-shaped patterns as well as incomplete circles, this visual language is a tribute to ongoing life. Scientists too isolate conspicuously corresponding patterns of fragments within fields when they use electron microscopes to image chromosomes (Fig. 16). Random, yet with decipherable meaning, these forms, like the Mbuti symbols, are a way to order and classify living organisms. Thus, for both cultures, similar figures move between notions of fact and fetish.

Examples of reduced sign notations from the prehistoric to the modern are intriguing to compare. Organizing charts of human figures (Fig. 17), as they appear on Spanish petroglyphs, archaeologists create a morphology of types.[40] Their arrangement allows the investigator to compare, contrast, and visually determine similarity and difference. In a karyotype chart (Fig. 18),[41] the laboratory technician arranges metaphase chromosomes into visual schemes based on size and number. In both cases—the petroglyphs and the karyotype charts—images operate at the intersection of word and picture. For the archaeologist, the quasi-human forms are prehistoric writing: for the geneticist, the karyotype is a chart representing the conversion of life form into sign language. For artists, such sign notations are sources of inspiration to be integrated and transformed in their work.

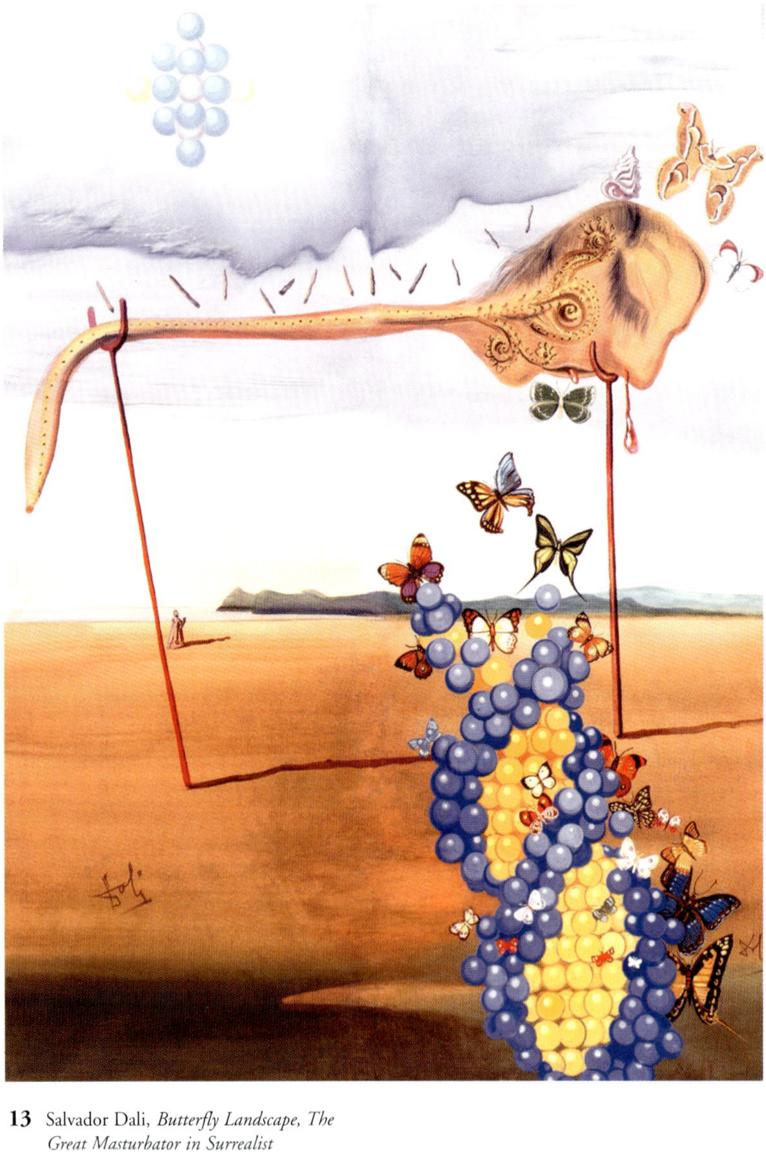

13 Salvador Dali, *Butterfly Landscape, The Great Masturbator in Surrealist Landscape with DNA*, 1957–1958.

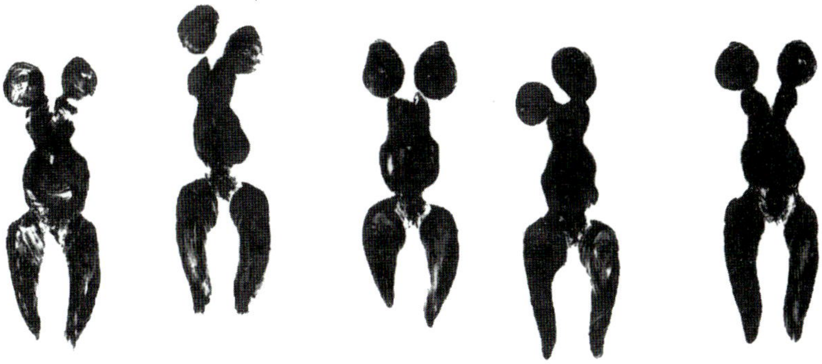

14 Yves Klein, *Anthropométrie de l'époque bleue (Anthropometry of the Blue Period)*, 1960.

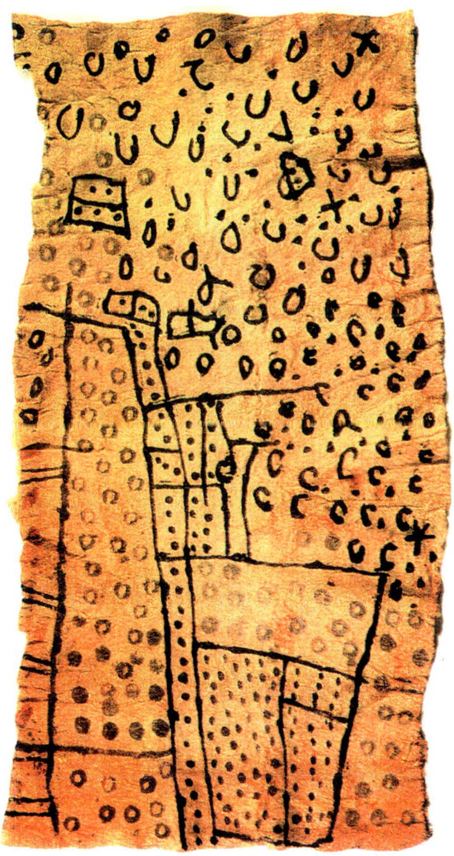

15 Zaire drawing, Mbuti tribal design, 20th century.

16 Micrograph of sex chromosomes.

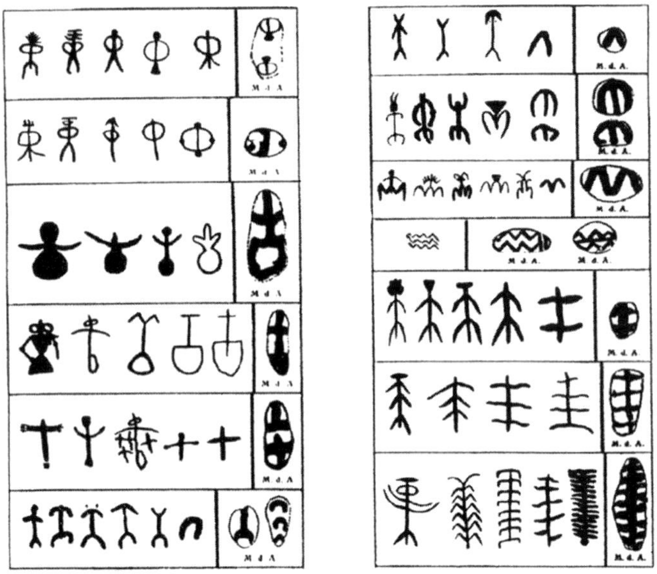

17 Conventionalized human figures from Spanish petroglyphs, compared with similar designs from the painted pebbles of Mas d'Azil, Ariege. After H. Obermaier.

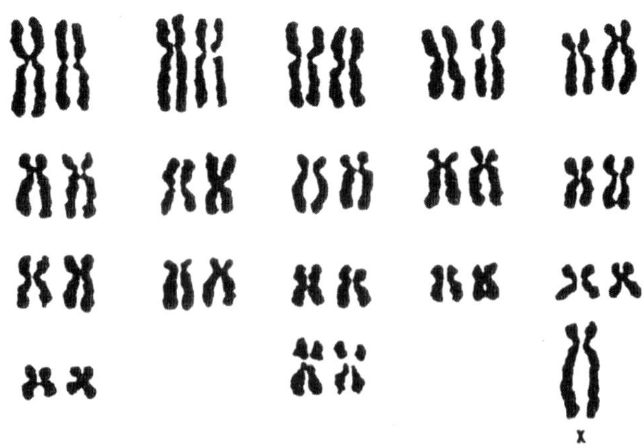

18 Karyotype of hammer-headed fruit bat.

An interest in language and its intersection with the body has been a frequent theme in the history of art. The language metaphors in visual art represented here as historical considerations belong to varied stylistic idioms,[42] but they help to illuminate the current interest in scientific images and their continuity with iconographies of the past.

Picturing DNA in Contemporary Art

The 1980s saw a renewed interest among artists in identity, the body, and its politics. Much-publicized findings of geneticists added a further dimension to this preoccupation. The claims of this growing field have engaged the imagination of many artists drawn to a science that appears to reveal the very essence of life. As molecular biology turns the body into text, high-magnification microscopes, mapping devices, and computer simulations generate novel ways to reveal biological information, accelerating its visual appeal. Perceived as language or a code, DNA has become, in effect, a molecular metaphor, a way to probe the reality behind appearances. As both a cultural and biological consideration, DNA is to many artists a magical molecule, a supermolecule. What other entity carries within itself a record of its own history? What other substance can self-replicate and carry information ad infinitum into the future?

Artists adopt the signs of semiotics to represent DNA's molecule in three distinct ways—as icon, as index, and as symbol.[43] The American pragmatist Charles Sanders Peirce formulated this nomenclature to describe the kinds of signs operative in linguistic structures. Reflecting the ascending interest in semiotics in the 20th century, art historians such as Rosalind Krauss,[44] W. J. T. Mitchell,[45] and Norman Bryson[46] invoke this typology to understand visual images. It is a useful and revealing way to decode linguistic metaphors in DNA art.

In an *icon*, the representation of the created image reflects the fidelity of the subject portrayed. The physical resemblance is self-evident as in a mirror image or a photograph. In representing genetic material, artists, taking their cue from geneticists, focus on different levels of cellular and molecular organization from genes to chromosomes to the helical structure. As genetic icons, both the double helix and metaphase chromosomes appear in painting, sculpture, photography, and video installations.

The double helix has become the 20th century's iconic molecule. Tony Cragg, in *Code Noah* (1988), welds together stuffed animals, cast in bronze, to construct a spiral chain. Consisting of two intersecting bands, the twisting sculpture employs the physical structure of the double helix (Fig. 19). By playfully building a molecular architecture with children's toy animals, Cragg conveys the connection between natural and cultural artifacts and suggests that nature itself is culturally constructed.[47] The double helix also

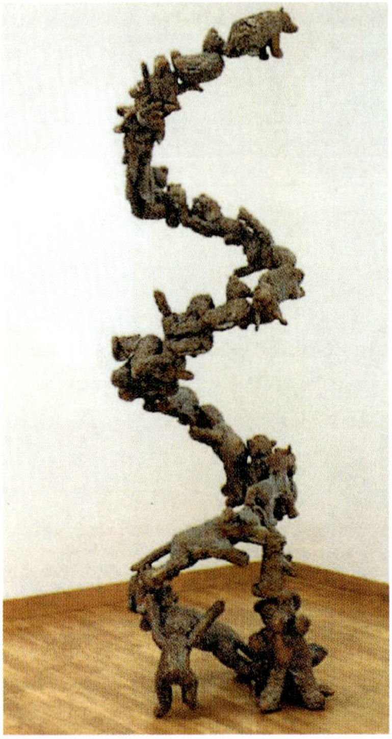

19 Tony Cragg, *Code Noah,* 1988.

appears in architecture and outdoor sculptures at universities and laboratories. At Cold Spring Harbor Laboratory, for example, architect Charles Jencks created *Spiral Time* (2000) on campus. The University of California commissioned sculptor Roger Berry to build *Portrait of a DNA Sequence* (1998) for the Life Sciences building at the Davis campus, and an elaborate G-Nome installation was created by William Leicester for the University of Iowa. The double helix ladder shows up in unexpected popular venues, such as a playground in South Beach in Staten Island, New York; sidewalk chalk figures; origami models; Lego constructions; jewelry; perfume bottles; body tattoos; and a variety of other ornaments bordering on kitsch.[48]

So too, chromosomes, from the Greek, meaning "colored bodies," have become iconic representations. Chromosomes, made up of DNA molecules and their attendant proteins, carry the sequence of genes. Located in the nucleus of the cell, these thread-like structures are readily visible under the microscope during metaphase—that part of the cell cycle in which they elongate and replicate into identical pairs. Chromosomes appear in several sizes, shapes, and styles that are reminiscent of arcane lexicons. They provide a set of visual images that represent the instructional language of the body. To the artists who use these images, the body seems to write itself as a string of semiotic fragments in an information system.

The installation pieces of Suzanne Anker, for example, employ greatly enlarged chromosomes that look like ancient alphabets. Interpreted by art historians Barbara Maria Stafford and Frances Terpack, these pieces translate the genetic code into a universal language system, the lingua franca of life (Fig. 20). For Anker, the X shapes of the chromosomes become miniature bodies. Using optical abstractions and anthropomorphic figures, she "transforms linguistic grids and codes into playful cavorting dancers"[49] (Fig. 21).

Photographer Gary Schneider, in his series of genetic self-portraits comprising 55 silver gelatin prints, actually tags the gene sequences on the chromosomes.[50] He employs sophisticated light and electron-scanning microscopes as well as X-ray technology, creating images that resemble the chromosomes so clearly that they meld the distinction between art and scientific photography (Fig. 22).

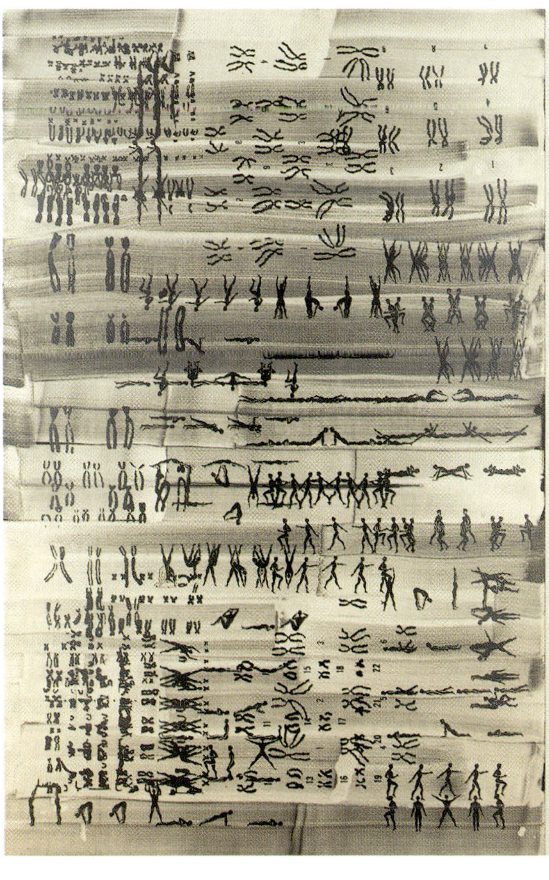

20 Suzanne Anker, *Micro Glyph (Soma Font)*, 2000. Collection: Goldblum Lévy, Switzerland.

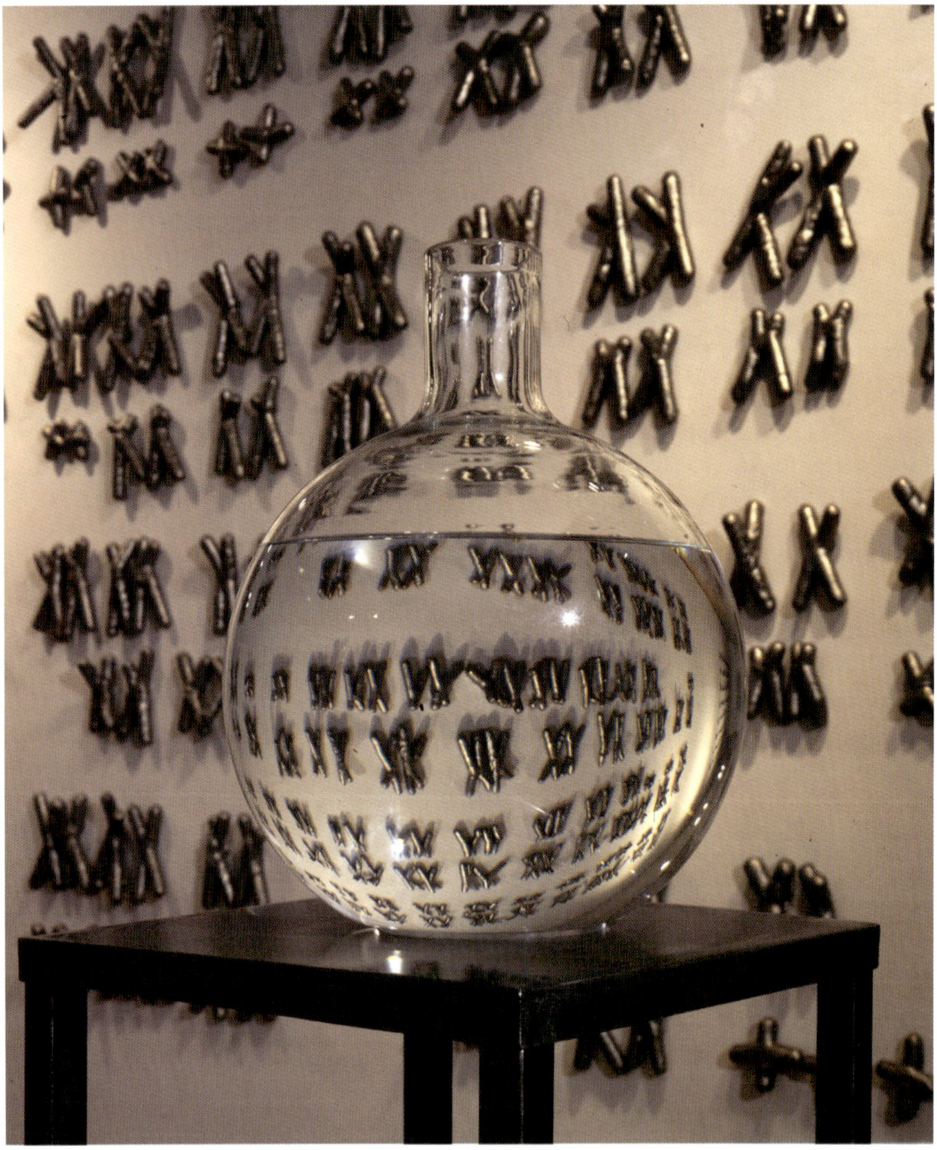

21 Suzanne Anker, *Zoosemiotics: Primates, Frog, Gazelle, Fish* (detail), 1993.

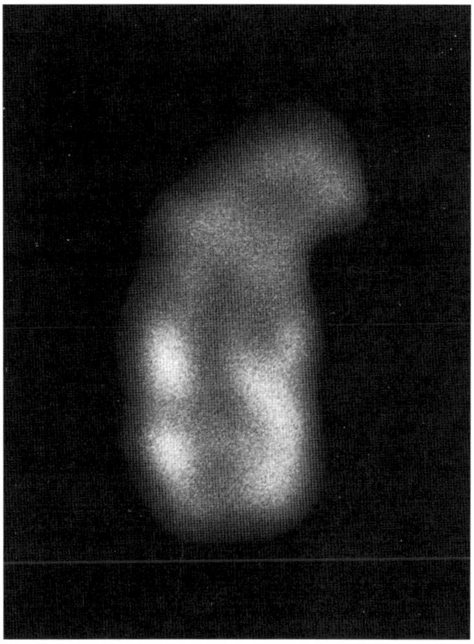

22 Gary Schneider, *Chromosomes* (from *Genetic Self Portrait),* 1997 (Detail/Y chromosome).

The DNA molecule is also represented in visual art as an *index*—a form that is produced indirectly in much the same way as a shadow is inflected by the sun. An autoradiograph is the bar code pattern chemically produced by an individual's genetic code. It is an index, a marker of personal identity—the DNA fingerprints that became so familiar as forensic evidence during the trial of O.J. Simpson. Autoradiographs consist of light and dark bands that form patterns in discrete rows. Such patterns are unique for each individual, and the technique allows investigators to match the DNA of a suspect to the DNA found at the scene of a crime. If the samples match, this provides an accurate means of identification. Thus, autoradiographs are widely used as the "gold standard" of evidence in criminal investigations, in cases of disputed paternity, and in controversial inheritance claims. But as reconfigured by artists, the autoradiograph can have multiple meanings.

Dennis Ashbaugh, in *Designer Gene* (1992) (Fig. 23), was among the first artists to employ these marking patterns in paintings. His large-scale paintings of autoradiographs employ light and color reminiscent of Mark Rothko's transcendental vision. However, his message is derived from the ability of the technology to translate a hidden reality into a visible pattern, to reveal the inner code beneath appearances. Ashbaugh is, in part, speaking

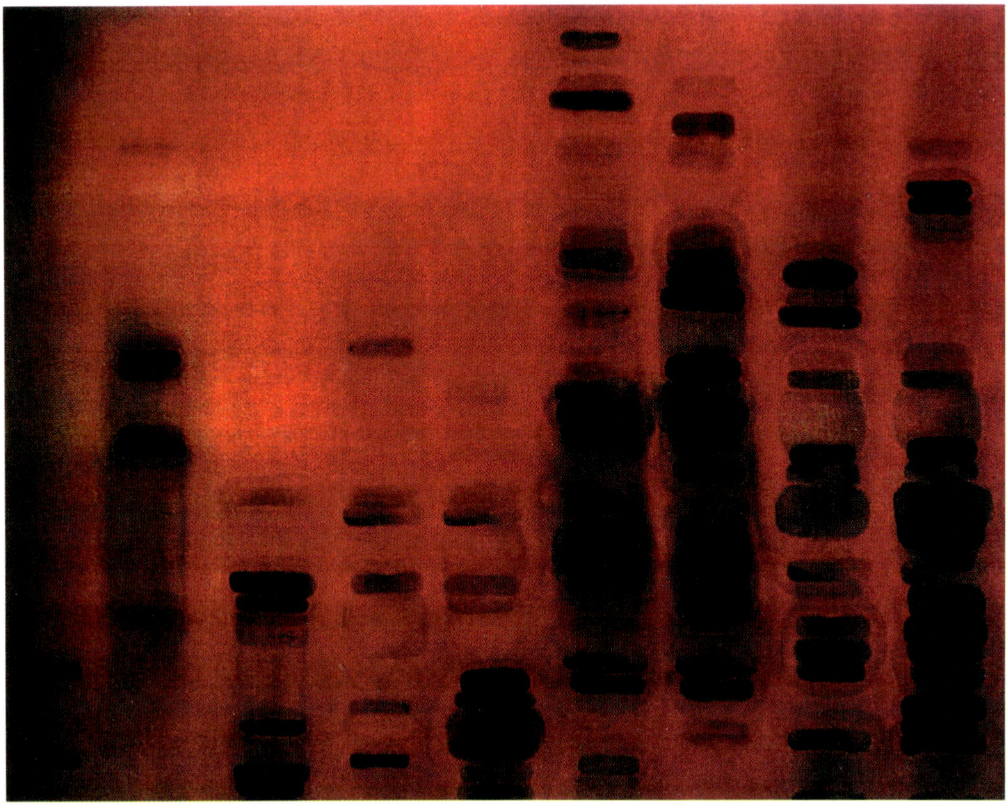

23 Dennis Ashbaugh, *Designer Gene*, 1992.

to his profession. He intends his work to create a new dialogue with color-field painting, in effect breaking the self-referential conventions of formalist art espoused by the influential art critic Clement Greenberg.[51]

The DNA portraits by Spanish artist Iñigo Manglano-Ovalle are more distinctly ideological. He embellishes autoradiographs through computerized color enhancement to make an ironical statement about the reductionism dominating the biological sciences. For example, when the wife of a patron of the arts wanted to give her husband a surprise portrait for his birthday, Ovalle devised an unusual way to capture the man's true likeness. He conspired with the patron's barber to pluck some hair from his customer's head and sent the sample to a forensic laboratory which extracted the DNA. The resulting bands were scanned into a computer. The result, *Clandestine Portrait*, appeared in a 1998 exhibition, "Garden of Delights" along with a collection of 48 other prints—all using genetic iconography as an index of individual biological differences (Fig. 24).[52] In other DNA portraits, mainly of his family and friends, Ovalle provides what he calls a post-racial view of identity by suggesting that the real identity of people lies not in skin color but in the invisible code beneath the skin—that is, in their DNA.

The meaning of a *symbol* is purely based on agreed-upon social conventions. Gene sequences, represented by ACGT arranged in a linear sequence, are not pictures of the actual molecules. They codify the sequence of nucleotides in each DNA molecule and thus are symbols that demonstrate the ways in which individual genomes differ from each other. The sequence of letters is distinct in the genome of each individual: We each possess our own genetic code.

This symbolic character of the genetic code is explicit in the genetic portraits created by Kevin Clarke. Clarke makes portraits using the letters ACGT to convey the idea that genes are the essence of personal identity. His DNA portraits strip away the individual's visual appearance and use instead that person's genetic code as a way to reveal his/her uniqueness. Clarke obtains the DNA sequence by analyzing the subject's blood through laboratory procedures. He then overlays the nucleotide sequence on an object that he associates with the person. For example, he superimposes James Watson's gene sequence on library shelves (Fig. 25), and he superimposes artist Jeff Koons' DNA sequence on a slot machine (Fig. 26). In these portraits, Clarke connects famous (or almost famous) people with the most famous of molecules. His technique calls to mind the devices used by some 19th-century romantic composers who based their melodies on notes derived from initials such as B.A.C.H. Today, composers are basing their compositions on the DNA sequence as a way to "tune in" to the dynamic of the cell cycle.

Steve Miller, one of the first artists to paint genetic portraits,[53] has recently turned his attention to the structure of proteins as in his *Eat Protein* series (Fig. 27). Consulting with Dr. Rod MacKinnon, a molecular biologist at Rockefeller University, Miller uses computer simulations to investigate the

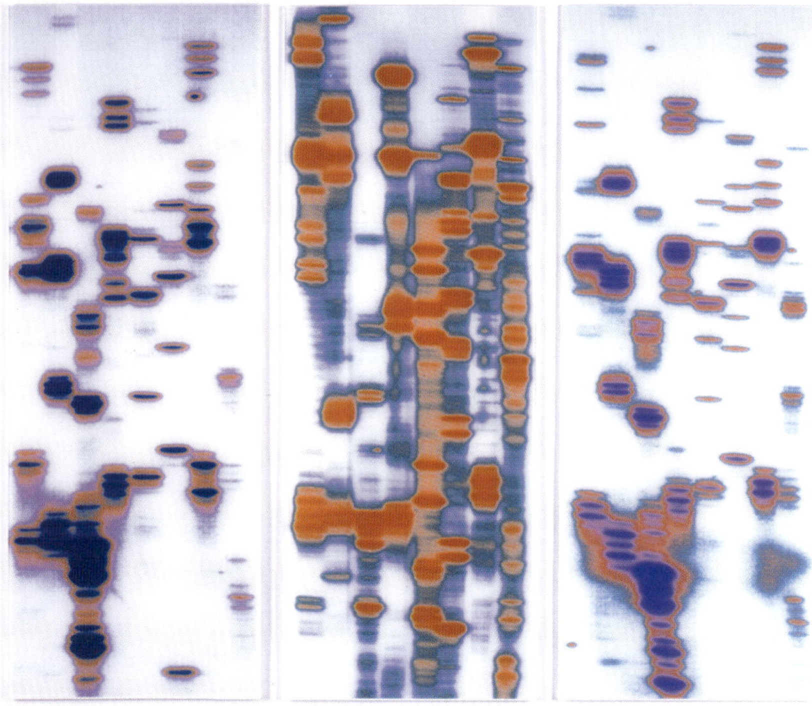

24 Iñigo Manglano-Ovalle, *Lu, Jack, and Carrie* (from *Garden of Delights*), 1998.

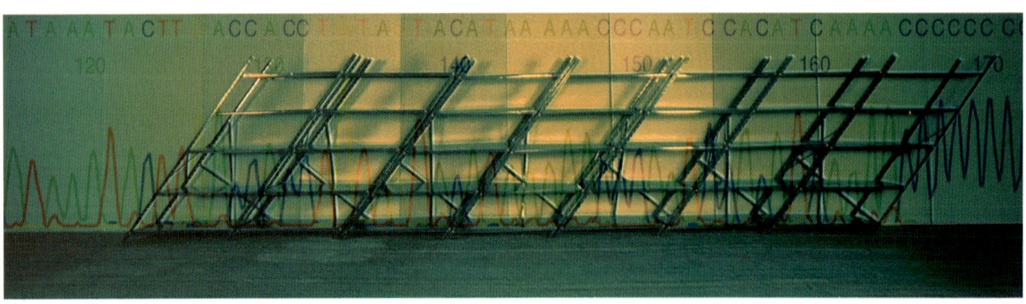

25 Kevin Clarke, *Portrait of James D. Watson* (panel 5), 1998–1999.

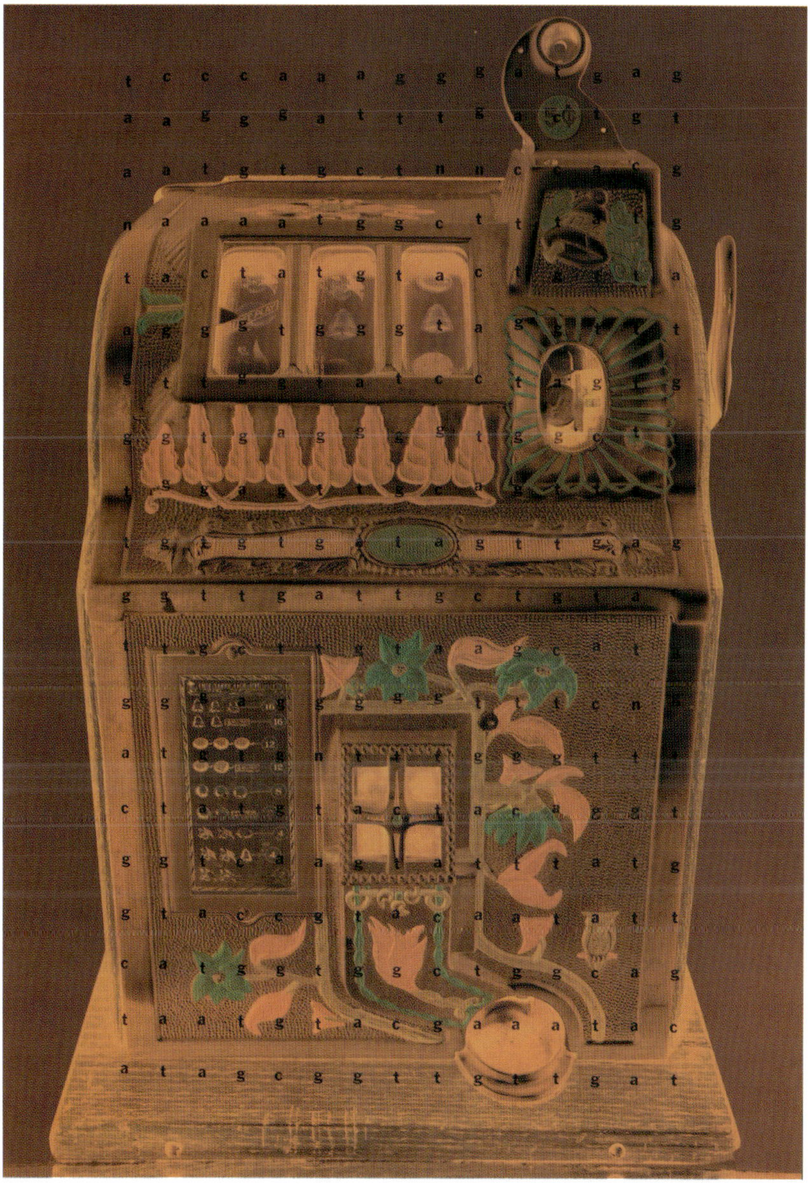

26 Kevin Clarke, *Portrait of Jeff Koons*, 1993.

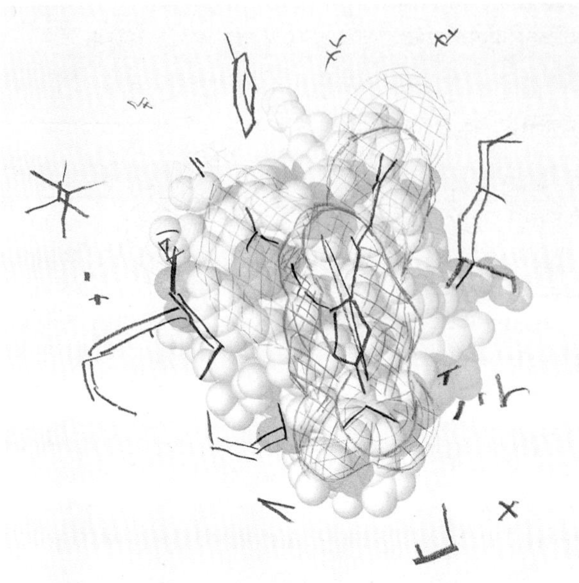

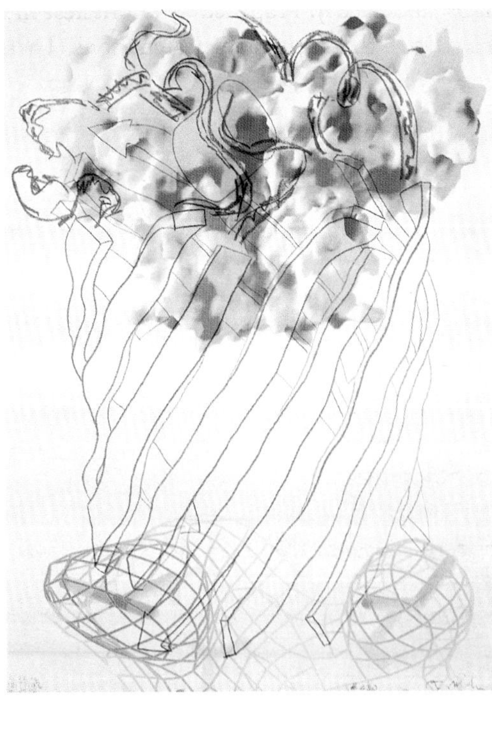

27 Steve Miller, *Eat Protein* (*top*) #230, 2000, (*bottom*) #259, 2000.

new scientific field of proteomics. This field, emerging from genomic studies, seeks to identify all the proteins expressed by a genome and their roles in physiology and pathology. In paintings and drawings, the artist portrays the three-dimensional structural patterns envisioned by this new molecular technology.

Marking Time: Language as Code

Other artists using the language of genetics in their work rely on more arbitrary or culturally specific signs. In the early 1990s Michael Joaquin Grey, an artist with a background in genetics, collaborated with Randolph Huff to create a computer program that generated proto-language. The resulting images, which appeared like the Rosetta Stone (Fig. 28), were inscribed on a slab of black graphite. Although driven by a number code, the forms themselves suggest the origins of both language and life.[54]

Photographer Olivia Parker in *Forms in Blue* (1999) (Fig. 29) superimposes code over a classic marble bust. Exposing information normally hidden from view represents for this artist the transformation of information itself and its impact on our lives. Choosing to discern what we wish to know based on technology is an attitude of mind of our information age. As a continuum between the ancient world and the present, code delivers to us an emergent pattern of being. Patterns of scientific imaging techniques also figure prominently in the work of Jane Lackey. Propelled by an interest in textiles as being representations of coded strings of concatenations, Lackey's painted wall objects employ variations of blots, smears, and successions of letter-forms in clusters and layers. These abstract layouts for the artist underscore nature's endless combinatory practice of recasting identity.[55]

In a large-scale, nine-panel color photograph set, Zhang Huan represents his *Family Tree* (2000) (Fig. 30). Using his own face as the ground or surface onto which inscriptions of family names are marked in calligraphic script, the artist draws attention to the ways in which identity is reconfigured generationally. By the end of the series, the obsessive print is unintelligible as writing but alludes to our common ancestry in black Africa.

Further forays of mapping information into the visual field are being conducted by Benjamin Fry in his *Genomic Cartography* series (2001) (Fig. 31). Displacing gene sequences with lines is a way to try to capture the movement in dynamic gene actions. Closer to a minimalist Agnes Martin painting than a Jackson Pollock, this work attests to the fascination that artists have with semiotic visual description.

Visualizing the Complex Structure of Life

Scientists visualize DNA exclusively through instrumentation—the medium of the microscope, crystallography, sequencing gels, magnetic imag-

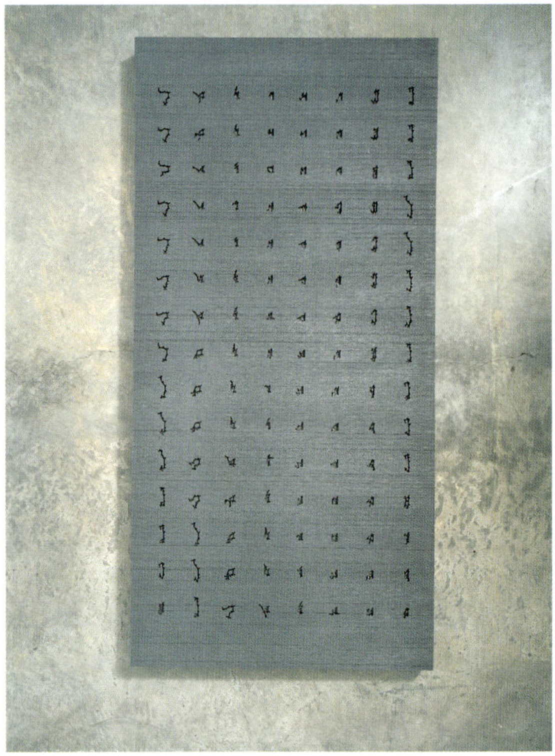

28 Michael Joaquin Grey and Randolf Huff, *Carbon Codex, 15 x 8*, 1992.

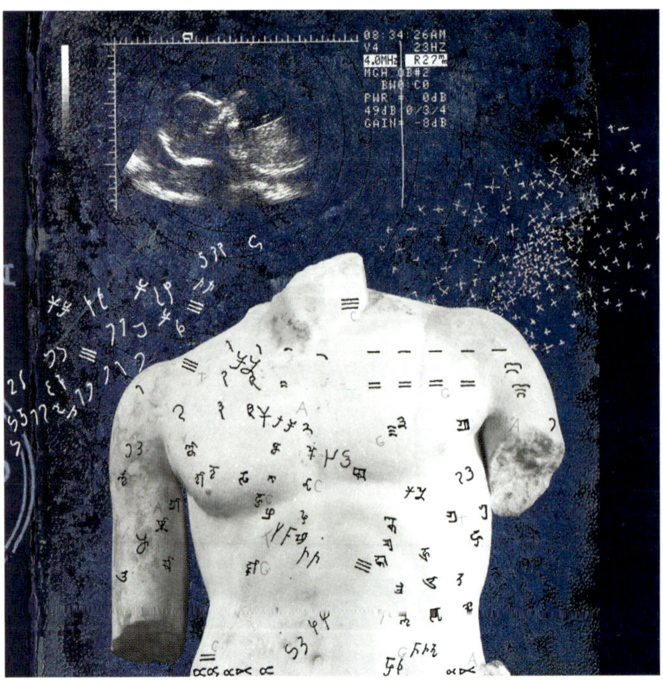

29 Olivia Parker, *Forms in Blue*, 1999.

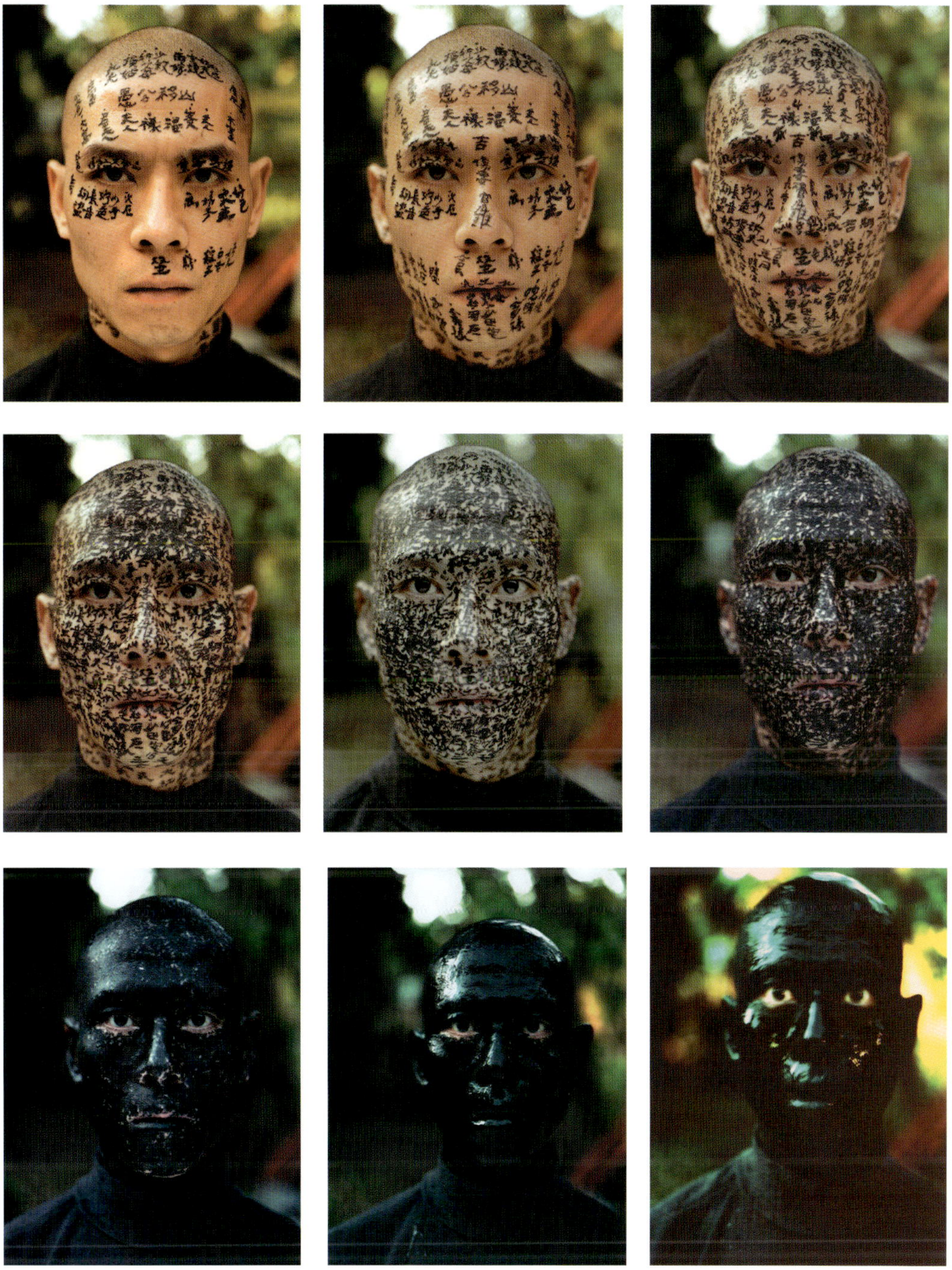

30 Zhang Huan, *Family Tree*, 2000.

31 Benjamin Fry, *Genomic Cartography: Chromosome 21*, 2001.

ing, and digitized images. Yet, these rather abstract visualizations of a biological entity are positioned to define the shimmering matrix of life. Interpreting DNA, avatar of 20th-century molecules, requires a certain blind faith in the link between image and actuality. Thus, it is not surprising that the molecular vision has been controversial among many scientists, scholars, and artists.[56] Reducing human life with all its social complexity to mere molecules, and assuming that all life's mysteries can be decoded, seems to denigrate the integrity of self as well as the social worth ascribed to human beings. Yet, gathering information through the deciphering of systems is the standard means of acquiring knowledge in virtually every field. Antony Gormley's *Bodies in Space IV* (2001) (Fig. 32), for example, reduces man's figurative presence to an array of walking molecules.

Reductionism may also obstruct scientific efforts to understand the body and its behavior. Writes historian Lily Kay, "Semantic value cannot be derived from technical notions of information...."[57] Anthropologist

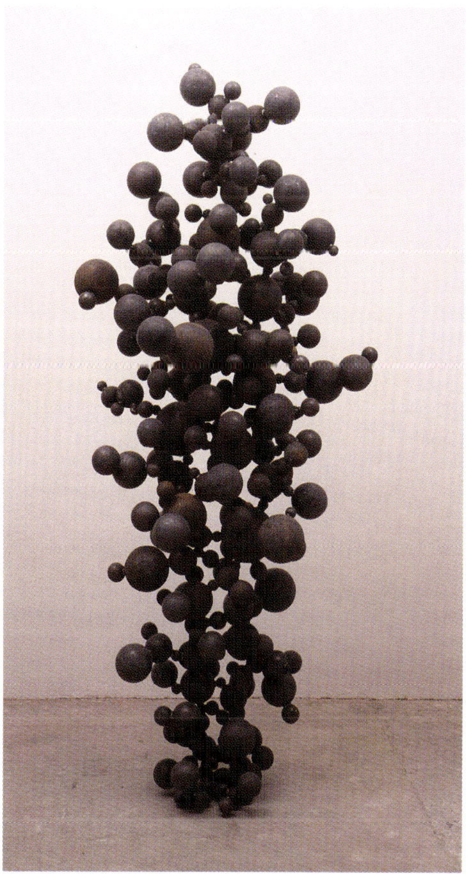

32 Antony Gormley, *Bodies in Space IV*, 2001.

Margaret Lock worries that the body has become "reified, isolated, decontextualized and abstracted from real time, actual location and social space."[58] And anthropologist Emily Martin writes about the disappearance of the body as we commonly perceive it.[59]

Most artists who work in this genre of art are less concerned with social critique than with visualizing the complex structure of life, but their own symbolic forms are open to wide interpretation. Because they reduce identity to a gene sequence—a code script of information—Kevin Clarke's or Marc Quinn's genetic portraits can be interpreted as too narrowly reductionist, or their work can be interpreted as an urge to reveal the inner domain of a person, concretized truth behind appearances. Ashbaugh's autoradiographs and Ovalle's DNA portraits can be seen as revealing the personal identity and invisible essence of their subjects, or they can be perceived as a satire on scientific reductionism. Additionally, the attraction to the signs and symbols of genetics can be viewed in its continuity with the language experiments of art in the past. Like the information in a genomic message, the meanings of this abstract and symbolic art must be considered in a more encompassing social and biological context. As Sir John Sulston commented about his genetic portrait in the National Gallery: "It's not me. It's my starting point."[60]

Notes

1. John Sulston and Georgia Ferry. *The common thread: A story of science, politics, ethics and the human genome.* Bantam Press, London, 2002.
2. National Portrait Gallery Press Release, Statement from Sitter and Artist 2001. Colin Martin. 2001. The scientists new clothes. *Lancet* 358: 1560; and "Gallery puts DNA in the frame," BBC News, September 19, 2001.
3. The most recent popular advocate of such reductionist claims has been Steven Pinker. In *The blank slate.* Viking, New York, 2002, he draws from evolutionary psychology and behavioral genetics to claim the biological basis of all human behavior. But reductionism has been widely criticized as ignoring the immense variation among individuals. See Evelyn Fox Keller, *Making sense of life: Explaining biological development with models, metaphors and machines*, Harvard University Press, Cambridge, 2002. Keller rejects the search for overarching principles, including wholistic theories. Reductionism is also criticized as based on ideological premises that justify the status quo. See R.C. Lewontin. *Biology as ideology.* Harper Perennial, New York, 1992. Also R.C. Lewontin, S. Rose and L. Kamin. *Not in our genes.* Pantheon, New York, 1984.
4. Charles J. Lumsen. 1986. The gene and the sign: Giving structure to post-modernity. *Semiotica* 62: 191–206.
5. Physiognomy, a term attributed to Aristotle, was popular in Paris in the first decade of the 19th century, appearing in widely circulated treatises by Swiss pastor Johann Casper Lavator (1741–1801). See Anthea Callen. *The spectacular body: Science, method and meaning in the work of Degas.* Yale University Press, New Haven, 1995, pp. 4–5.

6. Jennifer Terry and Jacqueline Urla, eds. *Deviant bodies: Critical perspectives on difference in science and popular culture.* Indiana University Press, Bloomington, 1981, p. 6.
7. Louis H. Marshall and Horace W. Magoun. *Discoveries in the human brain: Neuroscience, prehistory, brain structure, and function.* Brain Research Institute, Los Angeles, 1998.
8. Nicholas Mirzoeff. 1981. *Framed.* In Terry and Urla, op. cit., p. 67. The French still use the phrenological trope "Bosse des Maths" to describe someone with numerical skills. Thanks to Vera Zolberg for this information.
9. Sander Gilman. *Disease and representation: Images of illness from madness to AIDS.* Cornell University Press, Ithaca, 1988. Also Nicole Hahn Rafter. *Creating born criminals.* University of Illinois Press, Chicago, 1997.
10. Michael Gareau. *Charles le Brun: First painter to King Louis XIV.* Harry Abrams, New York, 1992.
11. Ibid., pp. 85–101.
12. Catalog Realize sous la Direction de Jean-Clair, L'Ame au Corps: Arts et Sciences 1793–1993, Galeries Nationale de Grand Palais, Paris, Octobre 1993–1994.
13. Eunice Lipton, *Looking into Degas: Uneasy images of women and modern life,* University of California Press, Berkeley, 1989.
14. Martin Kemp and Marina Wallace. *Spectacular bodies: The art and science of the human body from Leonardo to now.* Haywood Gallery, London, and University of California Press, 2000, p. 143.
15. Roger Cooter. *The cultural meaning of popular science: Phrenology and the organizing of consent in 19th century Britain.* Cambridge University Press, England, 1984.
16. Nicole Hahn Rafter. *Creating born criminals.* University of Illinois Press, Chicago, 1997.
17. Carol Squiers. "Perfecting Mankind: Eugenics and Photography." Exhibition at the International Center of Photography, New York City, January 11–March 18, 2001. The most influential source of these ideas was Charles Davenport. 1912. Eugenics and charity. *National Conference of Social Work* 39: 280–282.
18. Sander L. Gilman. *Difference and pathology: Stereotypes of sexuality, race and mad ness.* Cornell University Press, New York, 1985, p. 204. These ideas were prevalent at the time, appearing, for example, in the writings of Ernst Heckl and Sigmund Freud. In 1892, a physician, Max Nordau, author of a seminal text on degeneracy, extended this concept of degeneracy to include artifacts of culture including literature and visual art.
19. Stephanie Barron. *Degenerate art: The fate of the avant-garde in Nazi Germany.* Los Angeles County Museum and Harry N. Abrams, New York, 1991, p. 8. The German word *Entartet* is a biological term defining an animal that does not conform to the norm of its species.
20. Leon Poliakov. *The Aryan myth: A history of racial and nationalist ideas in Europe.* Basic Books, New York, 1974.
21. Erwin Schroedinger. *What is life?* Cambridge University Press, England, 1967. This work had significant influence on James D. Watson.
22. Claude Shannon. 1948. A mathematical theory of communication. *Bell System Technology J.* 27: 379–423, 623–656 developed the salient features of information theory, establishing the basic unit of information as the binary unit, and suggesting that information could be thought of as devoid of content. For a detailed history, see Lily Kay, who wrote *The book of life: A history of the genetic code.* Stanford University Press, Palo Alto, 2000.

23. For a review of this history, see Evelyn Fox Keller. 1990. *From secrets of life to secrets of death in body politics* (ed. Mary Jacobus, Evelyn Fox Keller, and Sally Shuttleworth), pp. 177–191. Routledge, New York. Evelyn Fox Keller, *The century of the gene.* Harvard University Press, Cambridge, 2000. Richard Doyle, *On beyond living: Rhetorical transformations in the life sciences.* Stanford University Press, California, 1997, p. 27.
24. Umberto Eco. *Semiotics and the philosophy of language.* University of Indiana Press, Bloomington, 1984, pp. 167–181.
25. See discussion in Jose van Dijck. 2000. The language and literature of life. In *Wild science* (ed. Janine Marchessault and Kim Sawanur), Chap. 5. Routledge, New York.
26. Robert Sinsheimer. *The book of life.* Addison-Wesley, Reading, Massachusetts, 1967.
27. Erwin Chargaff, cited in van Dijck, op. cit.
28. Robert Pollack. *Signs of life: The language and meaning of DNA.* Houghton Mifflin, New York, 1994, pp. 152–153.
29. Yoshimi Kawade. 1996. Molecular bio-semiotics: Molecules carry out semiosis in living systems. *Semiotica* 111: 195–215.
30. Bruce Wallace. *The search for the gene.* Ithaca, Cornell University Press, New York, 1992, p. 199.
31. Quoted in Leon Jaroff. 1989. The gene hunt. *Time Magazine*, March 20, pp. 62–71.
32. Joanna Drucker. *The alphabetic labyrinth: The letters in history and imagination.* Thames and Hudson, London, 1995.
33. Kirk Varnedoe. 1984. Gauguin. In *Primitivism in 20th century art: Affinity of the tribal and the modern* (ed. William Rubin). The Museum of Modern Art, New York, 1984, pp. 179–209.
34. Andre Breton. 2003. Manifesto of surrealism. In *Art in theory; 1900–1990: An anthology of changing ideas* (ed. Charles Harrison and Paul Wood), pp. 432–439. Blackwell, London.
35. Alfred Pacquiment, Henri Michaux, Centre Georges Pompidou, Musee National d'art Moderne, Paris, March 15–June 14, 1978, pp. 11–12, 70–71.
36. This reductionist language, called the Pitman System, was a common shorthand at the time. Andrew Robinson. *The story of writing: Alphabets, hieroglyphs and pictograms.* Thames and Hudson, London, 1995, p. 40. See also Suzanne Anker. "Writing Life's Gestures: Pollock and DNA," Lecture at the Pollock Krasner Study Center, Summer 2000. For text, see Ilene Roizman, Commentary. 2000. Pollack, DNA and language, *The East Hampton Star*, July 27.
37. See Harmke Kamminga and Soraya de Chadarevian. *Representations of the double helix.* Wellcome Unit for the History of Medicine, Cambridge, 1995, p. 31.
38. Nan Rosenthal. Assisted levitation: The art of Yves Klein. In *Yves Klein: 1928–1962: A retrospective.* Institute for the Arts, The Arts Publisher and Rice University, Houston, 1982. Although the author refers to these images as being both iconic and indexical, the bodies represented in this transfer are only partial representations, suggesting that their status as signs is intended as an index.
39. Georges Meurant and Robert Ferris Thompson. *Mbuti design: Paintings by Pigmy women of the Ituri Forest.* Thames and Hudson, London, 1995.
40. Sidra Stich, Joan Miro. *The development of a sign language.* Washington University Gallery of Art, St. Louis, March 19–April 27, 1980.
41. Although the species represented here is a fruit bat, all living forms, except perhaps the virus, are composed of these notational marks.
42. For art historian, Henry Focillon, "Iconography can be understood as the variation of forms upon the same meaning or the variation of meanings upon the same form."

Henry Focillon. *The life of forms in art.* Zone Books, New York. 1992. (Originally published in 1934.)

43. Floyd Merrell. *Peirce: Signs and meanings.* University of Toronto Press, Canada, 1997. For use of this typology in art, see Greg Myers. 1990. The double helix as icon. *Science as Culture* 9: 49–92.
44. Rosalind Krauss. 1977. *Notes on the index: Seventies art in America, Part II. October* 3 issue.
45. W. J. T. Mitchell. *Picture theory: Essays on verbal and visual representation.* University of Chicago Press, Illinois, 1994.
46. Norman Bryson. *Word and image: French painting of the ancien régime.* Cambridge University Press, England, 1981.
47. Suzanne Anker. 2000. Geneculture: Molecular metaphor in visual art. *Leonardo* 33: 371–375.
48. Soroya de Chadarevian and Harmke Kamminga. *Reproductions of the double helix.* Catalog of the Whipple Museum of the History of Science, University of Cambridge, England, 2002.
49. Barbara Maria Stafford. *Visual analogy: Consciousness as the art of connecting.* MIT Press, Cambridge, 1999, pp. 148–151.
50. Gary Schneider. *Genetic self-portrait.* Santa Barbara Museum of Art, California, 1997.
51. Clement Greenberg. 1986. Avant garde and kitsch (1939). In *The collected essays and criticism* (ed. John O'Brien). University of Chicago Press, Illinois, pp. 5–22.
52. Iñigo Manglano-Ovalle. The garden of delights. Exhibition at Southeastern Center of Contemporary Art, July 18–Sept. 30, 1998, Catalog.
53. See his "The Portrait of Isabelle Goldsmith." In Marvin Heiferman and Carole Kismaric. *Paradise now: Picturing the genetic revolution.* Tang Teaching Museum, Skidmore College, New York, 2002.
54. Richard Dorment. 1992. Drawing the future with a microchip. *The Telegraph, London.* September 16. The artist is also the inventor of Zoob, a modeling system and toy that emulates dynamic and living systems. Press release, Bitforms Gallery, New York, February 2003.
55. See Jane Lackey, In code, April 26 to June 15, 1996; essay by Laurie Palmer http://www.grandarts.com/JLackey.htm.
56. Marc Van Regenmortel and David Hull. *Promises and limits of reductionism in the biomedical sciences.* John Wiley & Sons, Chichester, 2002.
57. Lily Kay. *The molecular vision of life.* Oxford University Press, New York, 1993.
58. Margaret Lock. 1999. Genetic diversity and the politics of difference. *Chicago Kent Law Review* 75: 83.
59. Emily Martin. In Science and women's bodies: Forms of anthropological knowledge. Jacobus. op. cit., pp. 69–82.
60. Quoted by Martin, op. cit.

Mutation, Manipulation, and Monsters: The New Grotesque in Art

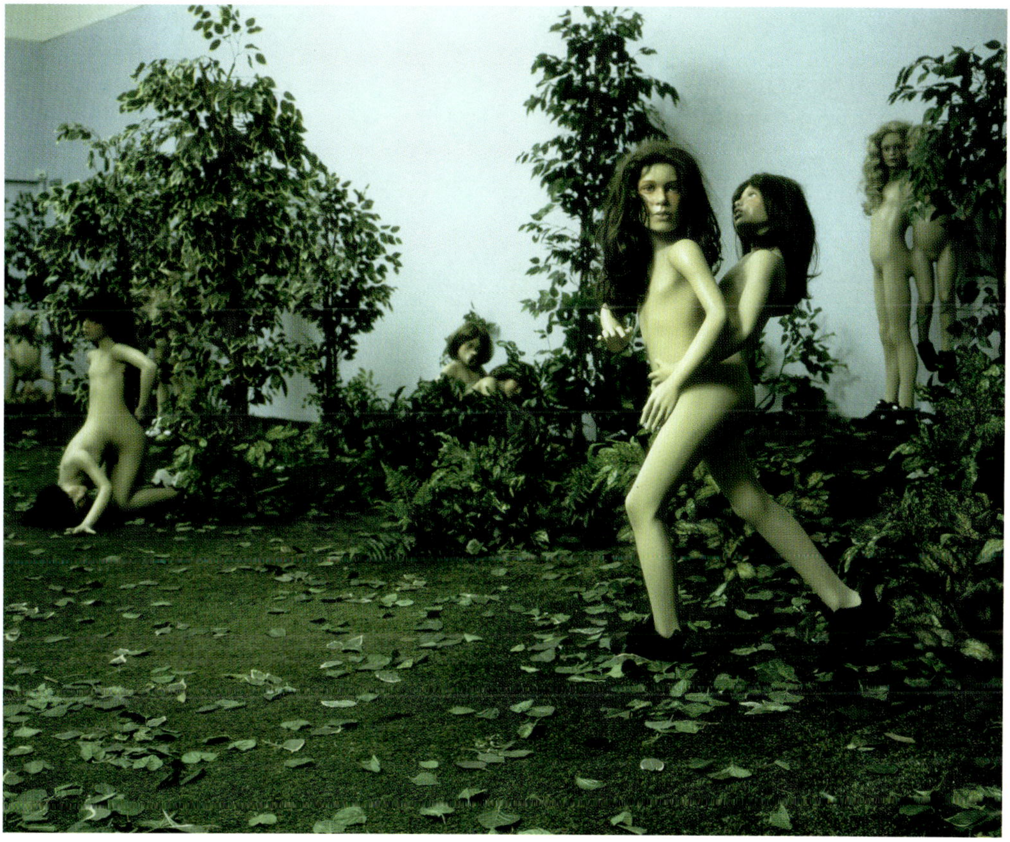

1 Jake and Dinos Chapman, *Tragic Anatomies* (detail), 1996.

THE CHAPMAN BROTHERS, in their 1996 installation, *Tragic Anatomies* (Fig. 1), portray conjoined twins, double-headed bodies, and a variety of prepubescent figures with gross malformations, all cavorting in a garden of the grotesque. These fiberglass sculptures are like factory-produced mannequins wrought by mechanical switches wildly out of control. Rude, impertinent, and invasive, these "in your face" creatures sport multiple sexual organs and orifices, in all the wrong places. Their installation recalls an old tradition of "monster-mongering" in circuses, carnivals, and side shows—displaying human oddities as monstrous marvels. Once called "freaks," today's "monsters" are construed as beings with mutations that are expressed as bodily aberrations. As genetics guides one's vision of the normal, anomalies have turned into congenital deformities, oddities into specimens, and monsters into mutations. And they unabashedly are portrayed as a "new grotesque" in contemporary art.

From Side Show to Science

In his classic 1978 book *Freaks*, historian Leslie Fiedler described the history of the cultural preoccupation with human oddities.[1] Fiedler explored the complex reactions—in life and in art—to anatomical mutations. He subtitled this book "Myths and Images of the Secret Self," for, he argued, we are intrigued by monsters as mirrors of ourselves. They reveal "for a moment out of time, the normality of a freak, the freakishness of the normal, the precariousness and absurdity of being, however we define it, fully human."[2]

Although those who are so different attract curiosity, they also evoke fear. The word "monster" is derived from the Latin verb *monere,* "to warn," suggesting that those who are different are dangerous. Associated with grotesque deformity, gross incongruity, and inhumane cruelty, monsters represent a threatening force, a terrifying and convincing deviation from the natural, but monsters are also associated with "marvels" because they seem to defy the rational order.[3] They evoke a sense of wonder, the sentiment that inspired the early explorers, naturalists, and collectors of unique specimens found around the world.[4] The grotesque, the anomaly, can also serve to define the boundaries of the normal and "natural." A deformed Caliban in Shakespeare's *The Tempest,* for example, compelled the audience to confront the meaning of the natural and to thus question the presumably orderly world of nature.[5] The great 17th-century Spanish painter Velázquez incorporated images of dwarfs in paintings of the Hapsburg Court. The dwarf in *Las Meninas* (1656) (Fig. 2), for example, is a man trapped in a child's body, a tragic figure, but his presence in the scene along with children defines his place in the society as a different kind of human being.

More frequently, human oddities and monster marvels were objects of public display. Popular side shows, traveling carnivals, and special exhibitions featured Siamese twins, midgets, giants, the extremely obese, and others with anatomical peculiarities. A famous and much-displayed rarity was the "two-headed boy of Bengal," whose head was conjoined at the crown with the head of his twin. The boy was born in 1783 to an impoverished peasant family in Bengal. His parents, in order to support the family, exhibited the child in Calcutta as a "monster." He died at the age of four from a cobra's bite, but his infamous double head lived on as an image both in clinical drawings and in visual art[6] (Fig. 3). Remarkably similar double-headed images appear in Bruce Nauman's enigmatic and psychologically terrorizing work, *Ten Heads Circle/Up and Down* (1990) (Fig. 4). Conjoined cast wax

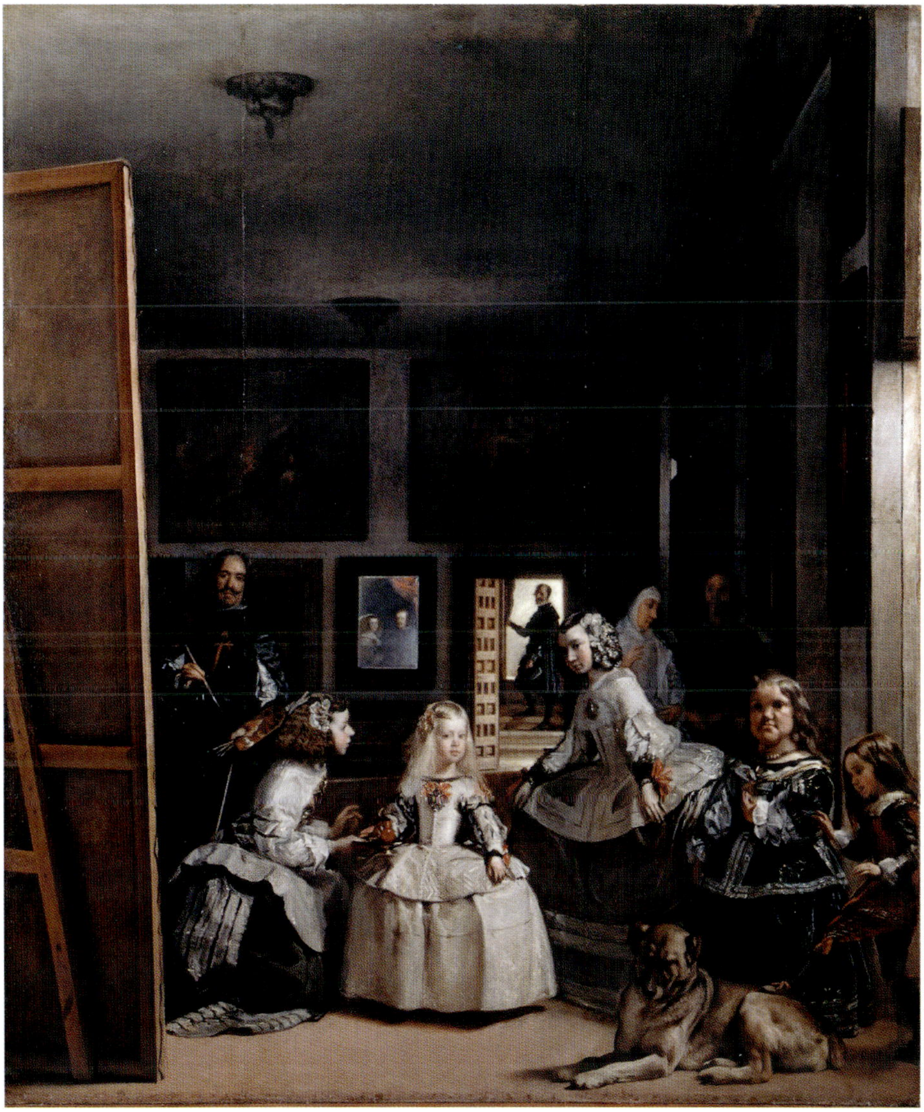

2 Diego Velázquez,
Las Meninas, 1656.

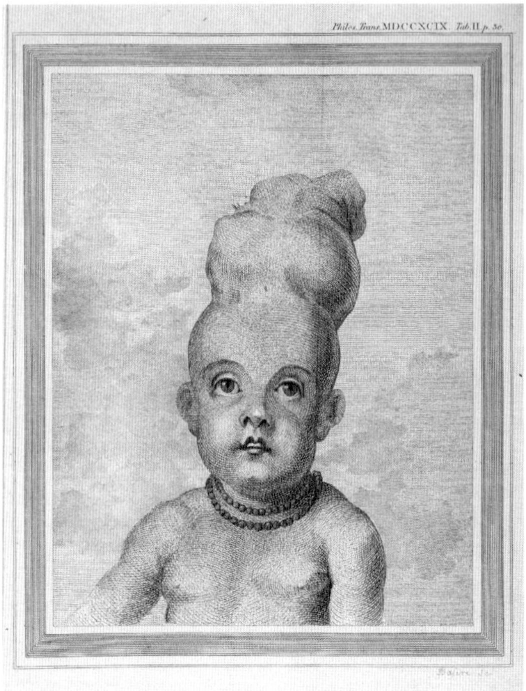

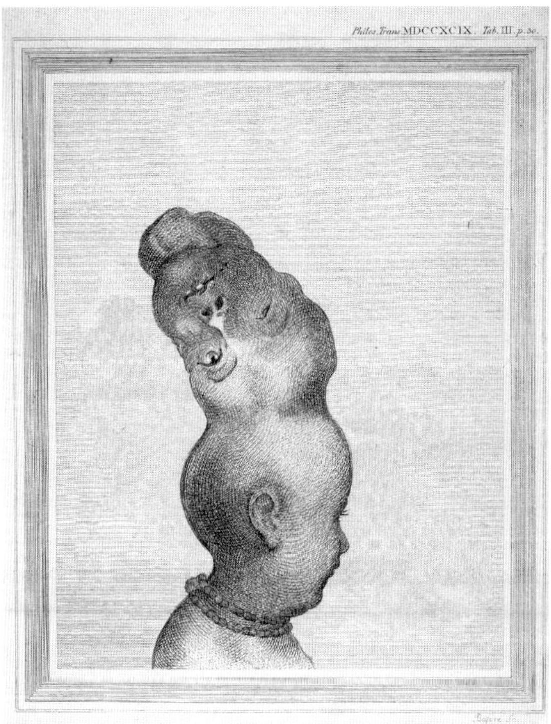

3 Mr. Dent, Plates II and III. Two watercolor drawings of the Two-Headed Boy of Bengal, by a Mr. Dent. Used by Everard Home in his 1799 paper on the boy. Engraved for the Philosophical Transactions of the Royal Society.

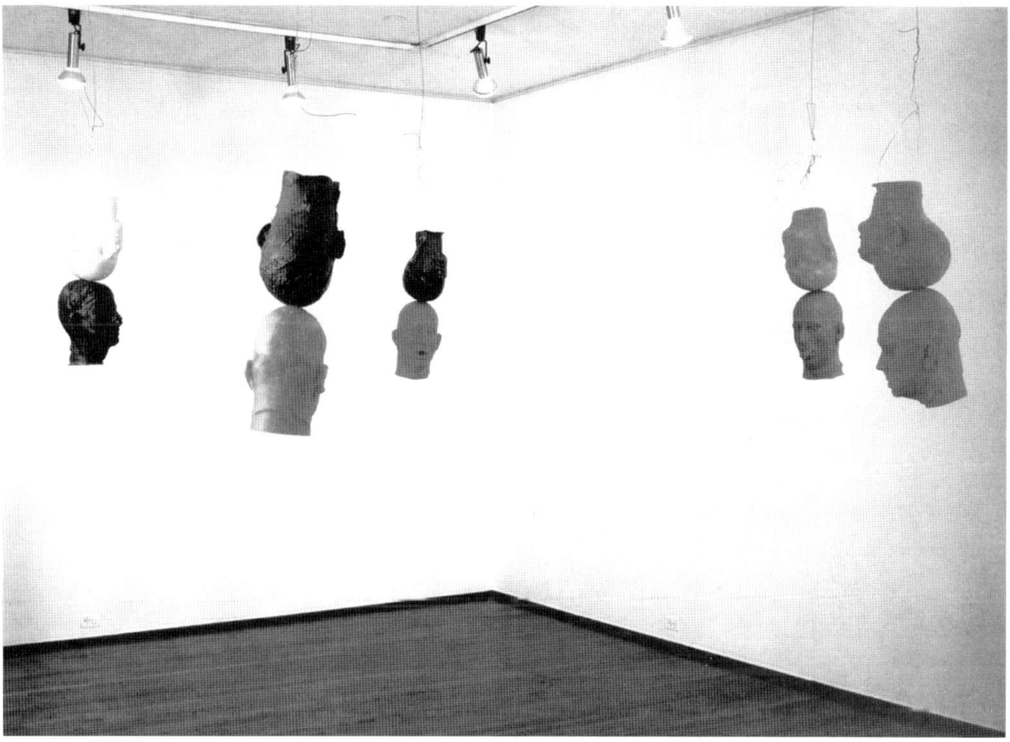

4 Bruce Nauman, *Ten Heads Circle/Up and Down*, 1990.

heads, suspended from dangling wires in space, confront us with a lurking second self, turning our world into paranoid shadows.

To study the "normal" in evolutionary life processes, medical scientists, in the light of the 18th-century phylogeny and ontogeny debates,[7] photographed monsters as ready-made specimens whose unique bodies harbored the secrets of biological development.[8] In the 18th century, the two heads of the boy of Bengal had been a clinical curiosity subject to empirical research and theoretical speculation about the nature of normality. By the 19th century, with the rise of photography, pictures of "special people" were marketed to the general public. Matthew Brady, the celebrated Civil War photographer, recorded images of performers in freak shows including, for example, *Barnum's Living Skeleton* (Fig. 5a) and *Barnum's Fat Lady* (Fig. 5b).

During the early years of the 20th century, scientists, influenced by Darwinian concepts of evolution, were deliberately trying to create monsters with experiments on living animals. Occasional experiments in monster-making began as early as 1774 with the production of multi-headed hydras. The term "teratology," the study of monstrosities or abnormal formations in animals and plants, was coined in the 1820s by Geoffroy Saint-Hilaire in his research of taxonomy and morphology.[9] A century later, in the 1920s,

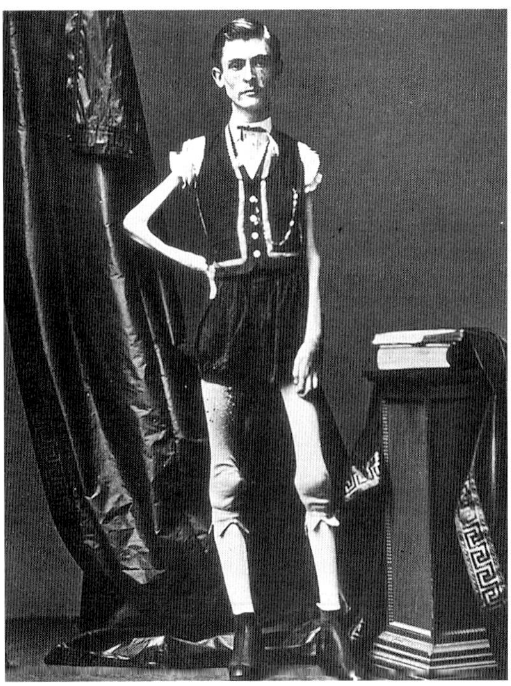

5a Matthew Brady, *Barnum's Living Skeleton.*

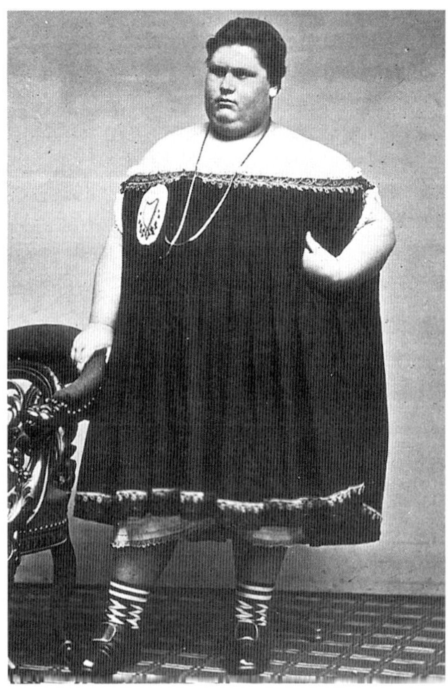

5b Matthew Brady, *Barnum's Fat Lady.*

German embryologist and Nobel Laureate Hans Spemann created two-headed tadpoles as part of his study of embryonic development. The art of monster-making expanded during the 1930s when teratologists began working with higher vertebrates and mammals.[10]

By the second World War, biologists were beginning to fashion creatures with congenital malformations by tampering with their eggs. Such experiments were then limited to fish, amphibians, and birds, but they were taking place just at that time when the atomic bomb was evoking anxieties about mutation and the biological effects of radioactive fallout. Such fears of fallout inspired a genre of monster novels and horror films about mutation.[11] They featured mutant creatures, as in Bela Lugosi's *Bride of the Monster* (1956), Lon Chaney's *The Atomic Monster* (1953), and a remarkable series of films in which ants, wasps, and scorpions grow into the size of 747s after straying into the path of atomic tests.

This concern about mutation persists as a theme in contemporary art. Cornelia Hesse-Honegger, formerly a science illustrator, turned to documenting flora and fauna surrounding nuclear power plant sites. *Four Seed Bugs,* Lygaeus equestris, *from Tubre and Santa Maria Madonna,* Italy (1994–1995) (Fig. 6) is a detailed watercolor of seed bugs with mutations evident in the pigment of their wings. Focusing on formal pattern arrangements in sequential order, the variegations in line and color point to malformations in these creatures. Hesse-Honegger also found seed bugs with similar mutations near Zion National Park in Utah, 240 miles away from an underground nuclear testing site.[12] Here, aesthetics becomes evidence, acting out a visual forensics at these sites.

Although focused on the dangerous effects of exposure to radiation, the preoccupation with mutation following World War II carried over to issues concerning the consequences of genetic research. In 1975, the very early recombinant DNA experiments at Harvard University inspired the flamboyant mayor of Cambridge, Alfred Vellucci, to alarm his community about the bizarre entities he called monsters that would crawl out of the laboratory. To an anxious public, such research, as well as any proposed change in the human constitution, was suspect: "Mutation meant monstrosity."[13]

Most mutation narratives follow an archetypal theme: Scientists attempting to change the attributes of human beings inevitably end up by being engulfed in disasters. Mary Shelley's classic *Frankenstein: or the Modern Prometheus* (1818) was, and remains today, the ubiquitous metaphor for attempts to create life by nonnatural means. The many stories of evil doctors—Dr. Frankenstein, Dr. Faustus, Dr. Moreau, Dr. Jekyll, Dr. Cyclops, Dr. Caligari—all convey the illusive ambivalent persona of science. The romantic view of the scientist as a "modern magician" parallels negative images, portrayed in hundreds of stories about scientists who are unable to direct or control the results of their experiments. Describing this public

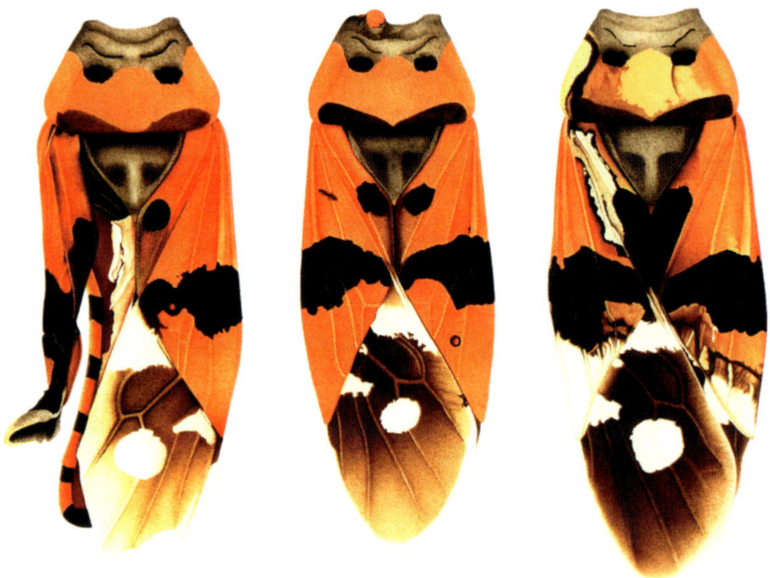

6 Cornelia Hesse-Honegger, *Four Seed Bugs,* Lygaeus equestris, *from Tubre and Santa Maria Madonna, Italy,* 1994–1995.

response to science, essayist Theodore Roszak wrote in 1974: "In the image of popular culture resides a legitimate public fear of the scientists' stripped down, depersonalized conception of knowledge—a fear that our scientists will go on being titans who create monsters."[14]

The side shows, with their focus on human abnormalities, were no longer in vogue during the latter half of the 20th century, reflecting the convergence of moral sanctions, legal prohibitions, scientific explanations of mutations, and broad changes in social attitudes toward those who are different. Collections of the anatomically grotesque were perpetuated, however, in "cabinets of curiosities," *Wunderkammern,* and displays at medical museums. Their evolution since the 17th century, and the revival of their popularity during the past decade, suggest the continued public attraction to the bizarre.[15]

Take, for example, the history of the Mutter Museum. Located at the College of Physicians in Philadelphia, this museum was founded in 1858 from the collection of Thomas Dent Mutter, a professor at Jefferson Medical College.[16] Exhibits include specimens of pathological tumors, diseased organs, and deformed skeletons. In addition, the museum features human oddities such as the connected livers of the famous Siamese twins Chang and Eng. Founded as a research and educational center, the museum reflected the hereditarian assumptions of the eugenics movement, and its collection served the pedagogical biases of medical sciences at that time. Medicine, then, was a predominantly descriptive science, so that body displays were essential teaching tools. After the turn of the century, studies of anatomy were replaced by more experimental fields of biochemistry and physiology, and scientific interest shifted from the structure of the body to its function. As its displays became peripheral to research, the Mutter Museum opened its doors to the public. In 1999 it loaned part of its collection, such as a 70-pound ovarian cyst, to a SoHo art gallery, Thread Waxing Space, for an exhibit entitled "Beyond Ars Medica." Visually compelling medical specimens, displayed in an art gallery context, attracted a significant audience curious about human oddities. Subsequently, the museum invited photographers to create an updated historical archive of "beauty, not in its conventional form, but in its opposite: the deformed, the broken, the disfigured body."[17] A photograph by Scott Lindgren, *Cephalothoracopagus* (1990) (Fig. 7), depicts a specimen of conjoined twins who had been delivered in 1852.

The National Museum of Health and Medicine at the Walter Reed Army Medical Center in Washington, D.C. has a similar history. The Museum was developed in 1862 by Surgeon General William Hammond, who collected medical specimens to help surgeons learn to treat battlefield wounds.[18] Such samples, preserved in formaldehyde, include bodies or skeletons of deformed infants and fetuses—a baby with two heads and an infant with a cycloptic eye—that had been archived at a time when eugenic

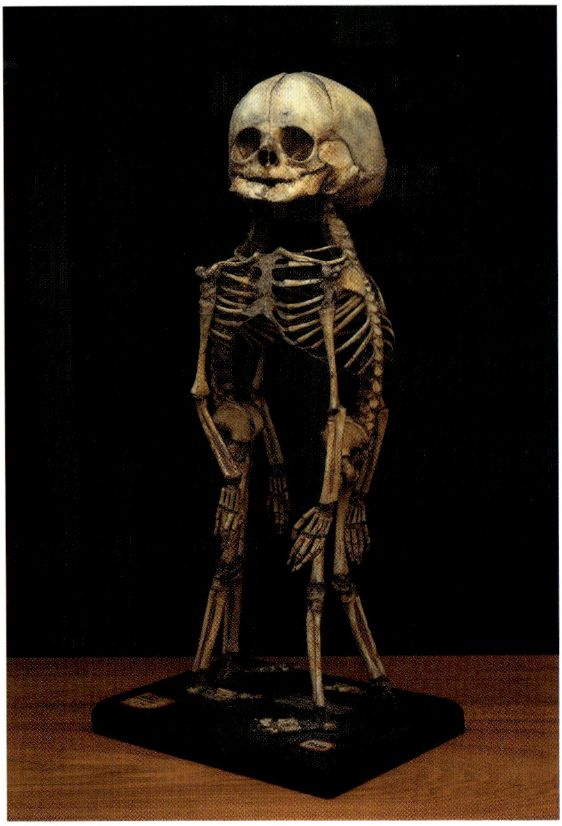

7 Scott Lindgren, *Cephalothoracopagus,* 1990.

interests led to a preoccupation with congenital birth defects. The scientific value of this museum and contributions to its collection declined as pathologists turned from an exclusive focus on whole organs to diagnosing disease at the cellular level. The medical museum currently displays preserved anatomical specimens, medical illustrations, interactive video installations, and images of the body gathered by advanced technological means.

In Amsterdam, the Museum Vrolik was established by Professor Gerard Vrolik (1755–1859) as a medical research and teaching museum. Since 1994, the museum has added 150 specimens of congenital malformations, including deformed fetuses ensconced in alcohol-filled jars. Orshi Drozdik, a Hungarian artist who lived in Amsterdam, has since 1983 photographed specimens in medical museums. Providing a rich source of visual "spike," medical museums continue to attract the curious. For full-fledged depictions of such, see http://www.rotten.com.

The Body and the Body Part: Switch and Bait in the Historical Archive

The grotesque is a complement of beauty, a blatant distortion or counterstatement of the aesthetic, and it has a long and fecund tradition in the history of art. Characterized by its early evocation of the intermingling of fantastic persons with foliage and plants or by bizarre incongruities and distortions of the body,[19] conceptual understandings of the grotesque, to this day, remain enigmatic.

In 1970, a conference of "The Grotesque in Arts and Literature" was held at the Whitney Museum of American Art,[20] bringing together writers, artists, curators, and theologians on the subject. Theologian James Luther Adams chaired the symposium. In his prepared statement, he cited the grotesque in art, the use of "monsters and the beastly... as depictions of frustrations and violence in contemporary existence." Whitney curator Robert Dory reaffirmed this position by stating "The artist desperately seeks to engage the mind and spirit of the spectator, to bring him into a state of awareness that will permit no evasion."[21] More recently, curator and writer Robert Storr is organizing an exhibition on "The New Grotesque" for the summer of 2004 at SITE Santa Fe, an international venue for contemporary art.

Artists during the late 19th and early 20th centuries projected surreal, even apocalyptic visions of monsters and mutations. Odilon Redon, in *The Cyclops* (ca. 1914) (Fig. 8), offers an example of this monster imagery as he responded to the theories of Darwinian evolution with fantastic images of mutating human and animal forms. The central figure in this phantasmagoric painting is a cycloptic, yet sympathetic, giant peering over a lush verdant landscape. In the *Odyssey*, Homer imagined one-eyed giants that he named Cyclops. Today cyclopia is defined as a genetic disorder called holoprosencephaly. For Redon, who had studied natural history, the image of the cyclops, which appears at various times throughout his work, was inspired by exhibitions of cycloptic lambs at natural history museums in Paris and Bordeaux.[22]

During the 1920s and 1930s, surrealism's affinity for the hyper-real brought together uncanny combinations of body parts. In her photo-collage, *Strange Beauty*, from the Ethnographic Museum Series (1929) (Fig. 9), Hannah Höch creates a disjunctive and bizarre combination as she draws attention to the ways in which cultural differences concerning beauty are expressed through specific bodies. The anomaly of a shrunken head placed on a western-style nude thwarts expectation. The Venus icon, a familiar symbol of feminine beauty, becomes unfamiliar, ugly, and strange.[23] So too, René Magritte, in *The Spirit of Geometry* (1937) (Fig. 10), creates an unlikely figure. An adult-sized baby in Madonna-like pose embraces a child's body on which the artist has transposed a mature head. By switching heads, the artist baits the viewer, eliciting an experience of the uncanny that invokes discomfort and anxiety if not illogical comprehension.[24]

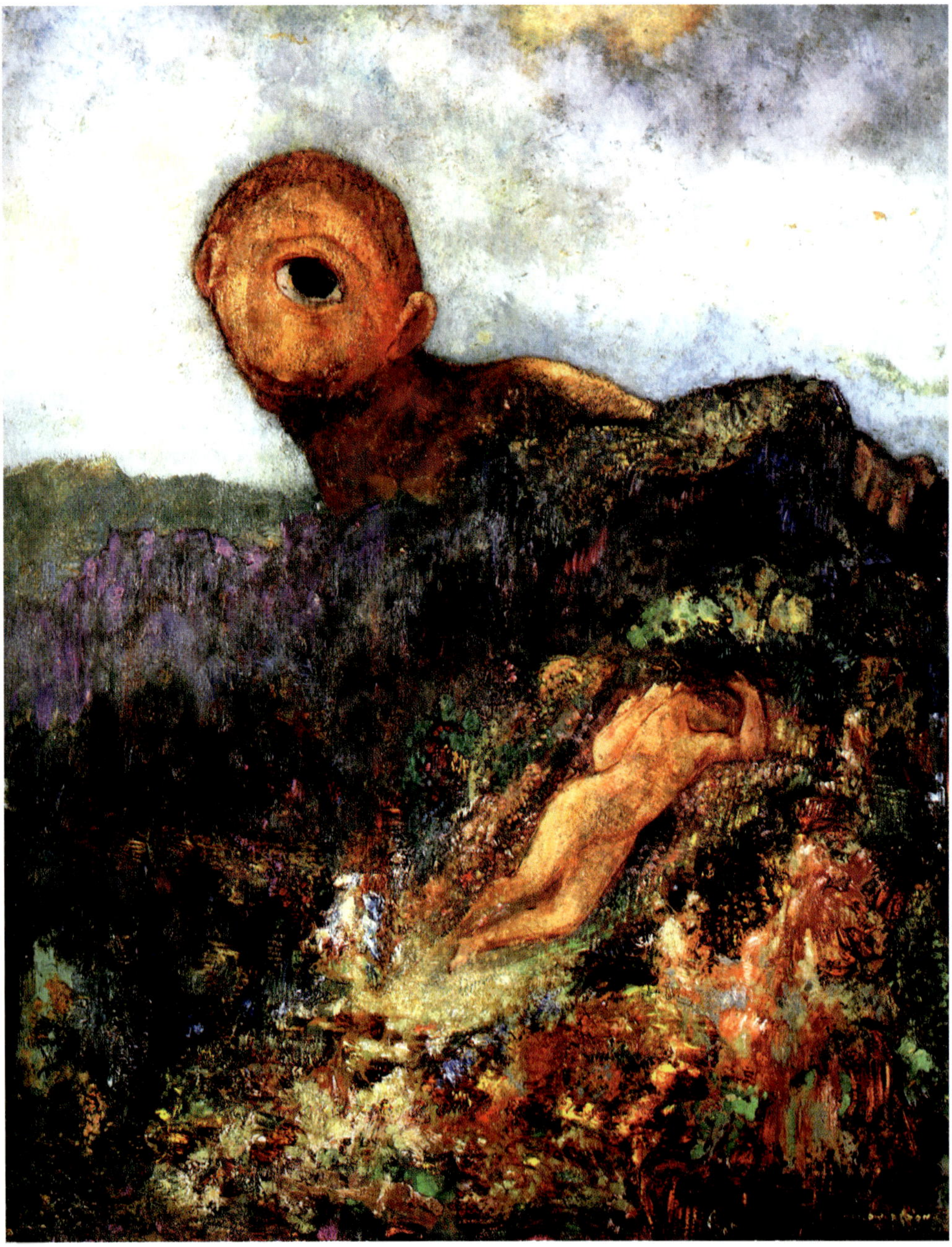

8 Odilon Redon, *The Cyclops*, ca. 1914.

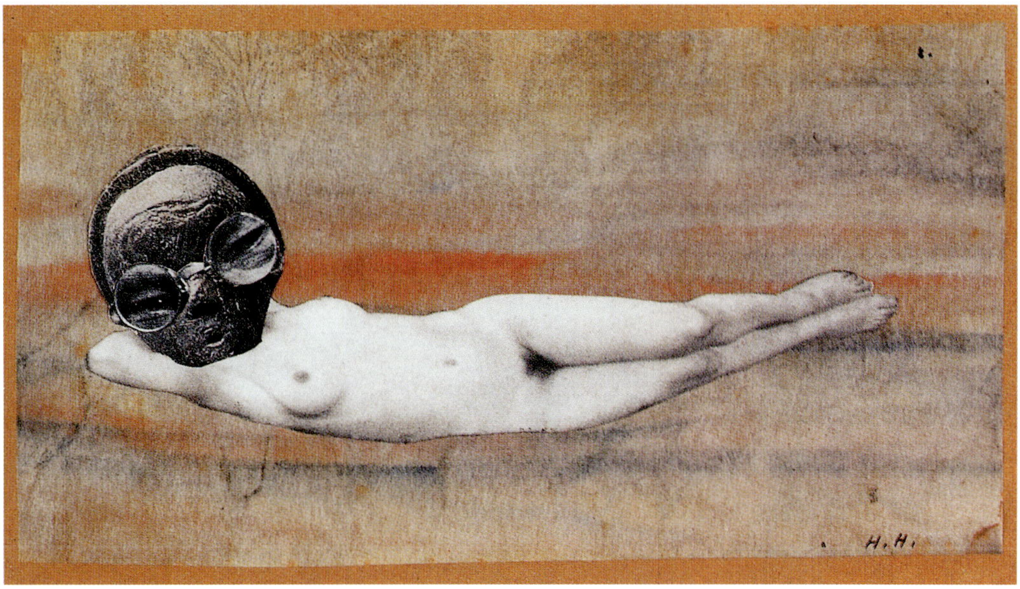

9 Hannah Höch, *Strange Beauty (Fremde Schoenheit)*, 1929.

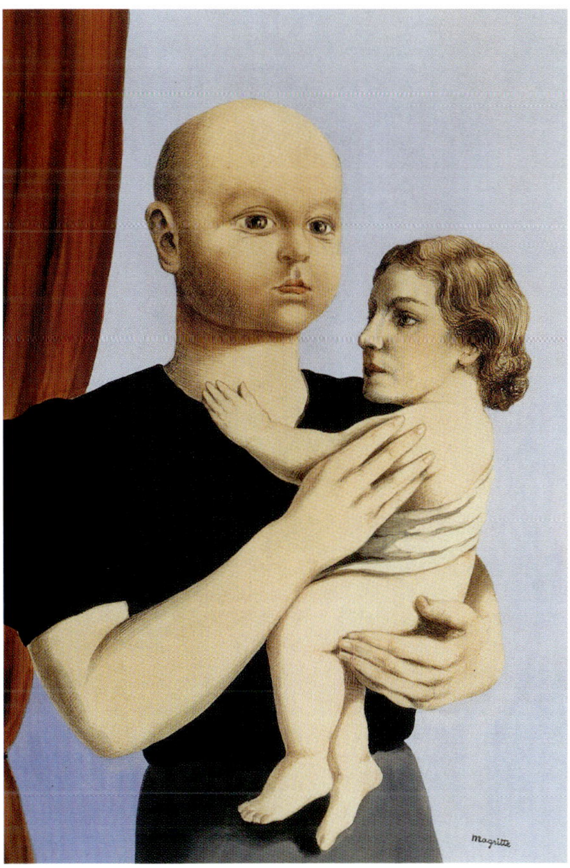

10 René Magritte, *L'Espirit de Géométrie (The Spirit of Geometry)*, 1937.

Hans Bellmer's fragmented doll bodies challenge the meaning and integrity of the normal. His photographs and sculptures of dismembered, distorted, and reconfigured mannequins are complicated and bizarre psychosexual icons. Bellmer compares the body "to a sentence that invites us to disarticulate it so that... its true contents may be re-composed." Accordingly, his art employs a process of disarticulation and recombination (the ersatz bodies were constructed with ball joints that permitted them to be fastened together). In *La Poupée* (*The Doll*) (ca. 1937–1938), a coy female character is casually poised on a chair (Fig. 11a). Dressed up with wig and bow, she confronts the viewer with bulbous breasts, swollen pubis, and absent arms. In an earlier variation (Fig. 11b), *La Poupée* (*The Doll*) wears stockings and boots and carries on her back multiple breasts which look like the yolk forms found in freshly gutted chickens. The deformed figures and deviant female bodies in the *Poupée* series can be interpreted as a countervailing commentary on the cult of the perfect body—the Aryan ideal of beauty that prevailed during the 1930s in Germany.[25]

Other artists have depicted transformations of the body and nature in response to advances in microscopy and the biological sciences.[26] Interest in the organic processes of the life cycle pervaded the aesthetic landscape in both Europe and America from the 1920s to the 1940s. Diverse artists at that time, including William Baziotes, Arshile Gorky, Roberto Matta, Theodore Roszak, and Mark Rothko, developed a pictorial style that has come to be known as biomorphism.[27] The paintings and sculptures produced by these artists represented cellular forms and organic allusions. They saw nature as an archive filled with odd remnants of a primordial past, and the body as a changing process of life, putrescence, and death. On the discoveries of modern science, writes art historian Jeffrey Weiss, "the painters grafted the primitive vantage of fear and awe."[28]

A Gruesome Parade of Horribles: Mannequins, Mutants, and Makeovers

In 1981, the United States Supreme Court, in a landmark decision, approved the patenting of a living organism.[29] The patent was for an oil-eating bacterium, but Chief Justice Warren Burger, in his decision, warned about the potential for "a gruesome parade of horribles." "We are told," he wrote, "that genetic research and related developments may result in a loss of genetic diversity and that its practice may tend to depreciate the value of life.... At times, human ingenuity seems unable to control fully the forces it creates—that, with Hamlet, it is sometimes better 'to bear those ills we have than fly to others that we know not of.'"

Justice Burger's warning addressed three concerns: Genetic engineering may lead to a parade of horribles—that is, monsters and mutations; genetic

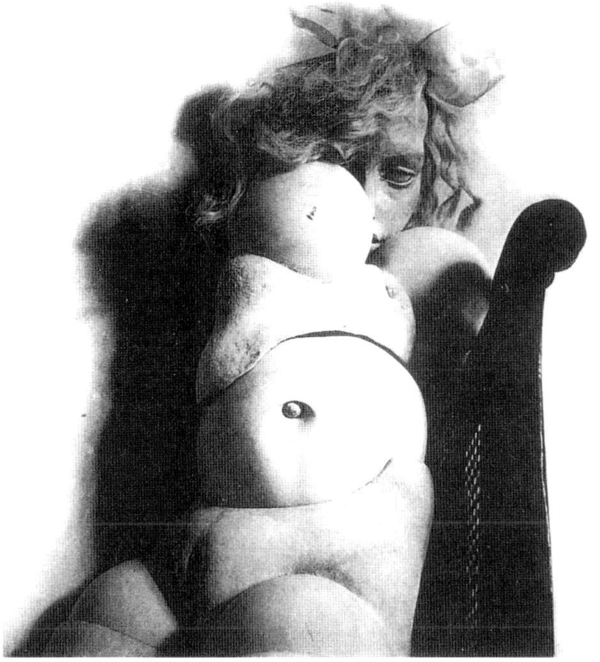

11a Hans Bellmer, *La Poupée* (*The Doll*) (detail), ca. 1937–1938.

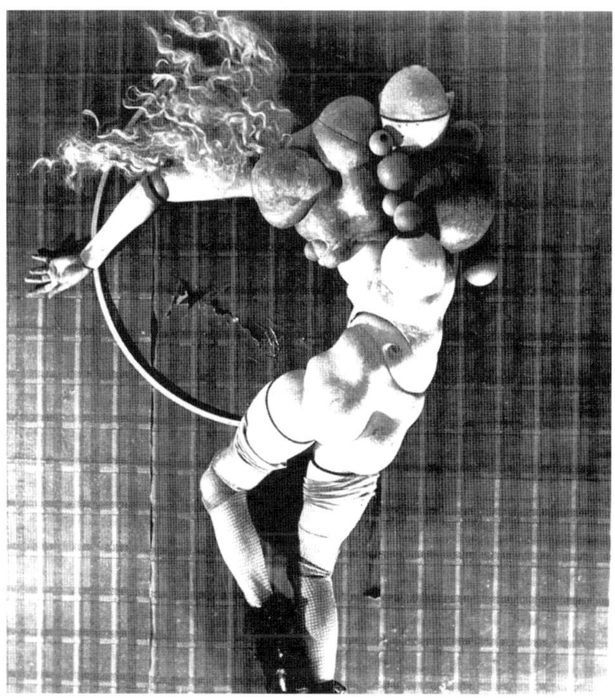

11b Hans Bellmer, *La Poupée (The Doll)* (detail), ca. 1937–1938.

research and related developments may affect the value of life and our concepts of normality; and biotechnology applications will have unpredictable social implications. These are among the salient issues reflected in contemporary visual art.

Apocalyptic Visions of Genetic Engineering

In popular culture narratives—Frankenstein stories, science fiction films, and mutation myths—DNA is forbidden territory to be transgressed at an exceedingly high cost, while genetic manipulation is taboo. Genetically engineered creatures inhabit the fantasy world of science fiction comic books and horror films. A series called *DNAgents* features creatures that act human, but whose DNA has been altered just enough to make them more than human, thus turning them into perfect special agents. In another comic book series, *X-Men,* a campaign is organized to "terminate genetic deviants." In yet another series, Dr. Xavier pleads with the public to "accept the muties" for "we are all related, we are all family."[30] Mutants also inhabit *The Fly* movies, both the original 1958 version and the 1986 Cronenberg remake. Whereas the first version evoked the fear of radiation, the remake emphasized the horrors of tampering with DNA.

Artists too use images of monsters to convey apocalyptic visions, and many of them are deliberately offensive—intended to shock. John Isaacs creates a monster by turning human flesh into a blob. His *The Matrix of Amnesia* (1997) (Fig. 12) depicts a corpulent human being as if he were an amoeboid organism. Acutely obese, and lacking a skeletal support structure strong enough to keep the body erect, the figure becomes a life-form reminiscent of primitive cellular development. Through excess, the monstrous and repulsive body regresses to an early evolutionary level, whereby protoplasmic creep becomes its form of locomotion. Isaac's freak defies human boundaries and magnifies the fragility of our concept of the normal.

So too do Rona Pondick's stainless steel sculptures, which transform the human body into a metallic skin. Shiny and flawless, emotionally distant, her works employ the synthetic process of assemblage. In *Untitled Animal* (1999–2001) (Fig. 13), a human figure lies insouciantly on the floor, apparently dissolving into body fragments. Like Kafka's beetle (or John Isaac's blob), the human form becomes a primordial monstrosity. Despite its shimmering metallic skin, the sculpture conversely conveys a dark sense of dissolution.[31]

In *Zygotic Acceleration, biogenetic, de-sublimated libidinal model (enlarged x 1000)* (1995) (Fig. 14), the Chapman brothers create an orgy of flesh by conjoining a dozen or more multiple bodies. The mannequins are positioned on a pedestal, recalling the 19th-century practice of putting "others" on display in museums and carnivals. Composed of connecting torsos with misplaced genitals, these transsexualized figures, writes a reviewer, "test the

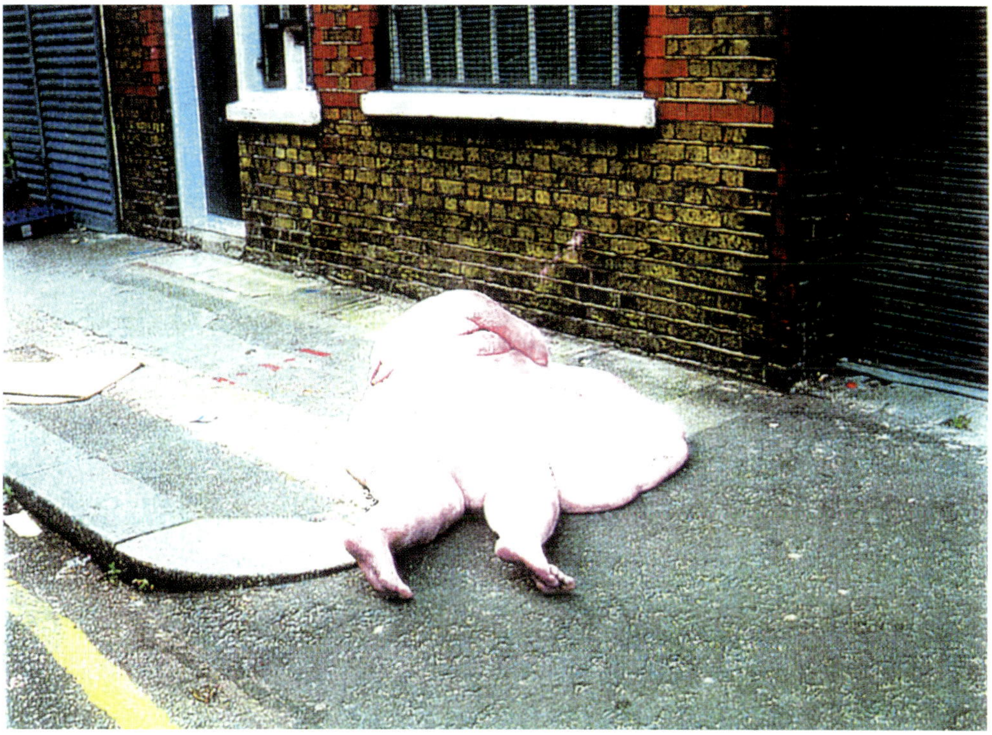

12 John Isaacs, *The Matrix of Amnesia,* 1997.

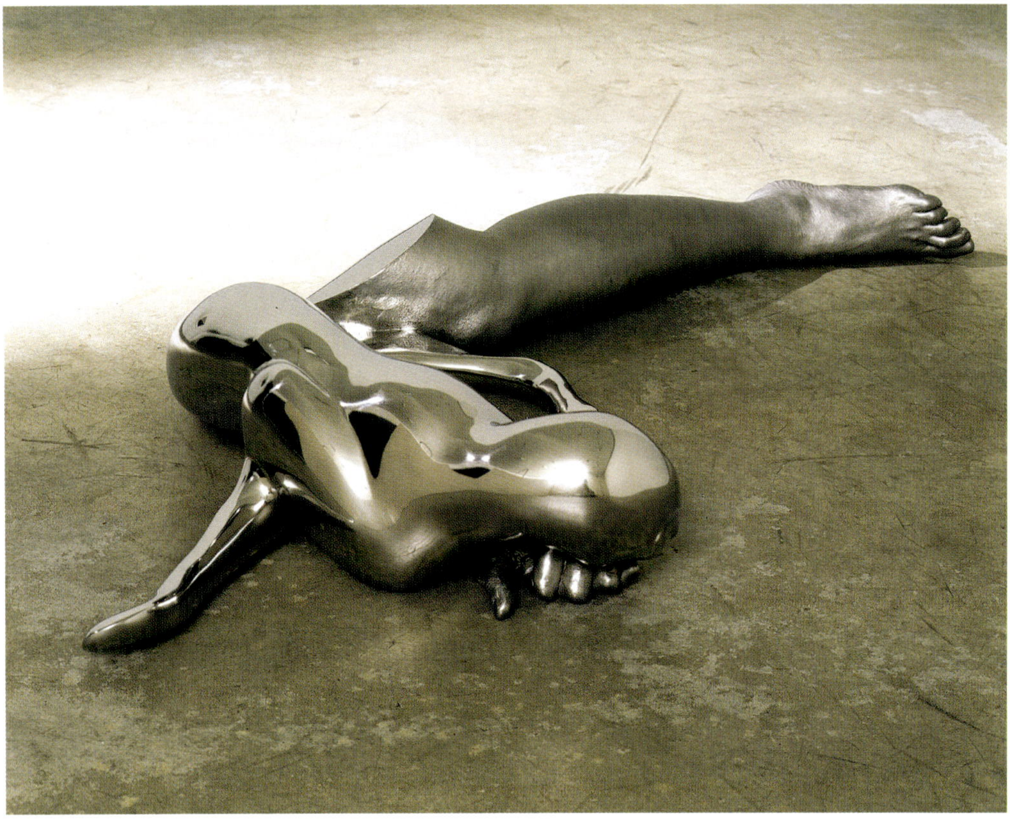

13 Rona Pondick, *Untitled Animal*, 1999–2001.

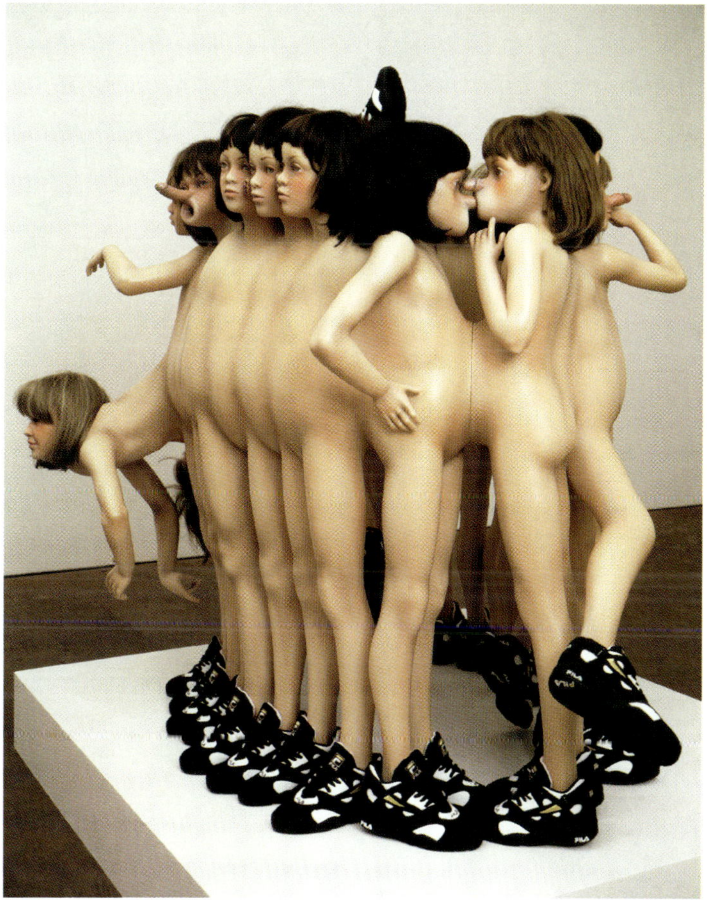

14 Jake and Dinos Chapman, *Zygotic Acceleration, biogenetic, de-sublimated libidinal model (enlarged × 1000)*, 1995.

limits of the repulsive and ridiculous."[32] In its carnivalesque presentation, the sculpture is rendered outrageous and absurd.

Cindy Sherman, the elliptical protagonist in her own photographic work, turned her attention in 1991 to mannequins as body doubles. Taking a cue from Hans Bellmer, Sherman elevates dolls to surrogates of corporeal beings which act out scenarios unthinkable in real life. In *Untitled* (1992) (Fig. 15a), Sherman comments with tongue in cheek on art history's representation of the Venus figure as an object of the gaze. Venus is frightfully dismembered, but with one last grasp of a recalcitrant stare, the observer is inescapably brought into the scene of the crime. In other works (not pictured here), the artist depicts a hermaphroditic mannequin shamelessly complete with penis and vagina. Presenting onlookers with reassembled prosthetic devices, she captures the projections of sexual fear, desire, and violence in her photographs of stand-in monster beings. Voyeuristic viewers may be terrorized and shocked, but also pleasurably enthralled.[33]

Fragile Concepts of the Normal

Holding the power to radically alter the cellular structure of human beings, genetic engineering is changing the very concept of what is normal, transforming our perceptions of the body, and, as Justice Burger warned in his U.S. Supreme Court decision, devaluing life. Genetic engineers can manipulate hereditary material through artificial selection, hybridization, or mutagenesis. They can decode, mutate, and clone specific sequences of DNA to produce organisms, cells, and subcellular components. Their purpose, like that of early teratologists, is to both grasp and influence the course of evolution by introducing desired genetic variation. Such techniques have a long history of agricultural and horticultural applications. Artists such as Edward Steichen and, more recently, George Gessert[34] (Fig. 15b), moved these horticultural experiments into art's domain. Gessert's hybrid flowers are "fictional" genomes bred for specific aesthetic characteristics. He uses terms such as beautiful, extravagant, boring, shy, vigorous, and strange to describe their characteristics.

Related techniques are beginning to be used in medical experiments for the replacement of defective genes. More controversially, however, genetic manipulation can also be used as a cosmetic procedure, allowing the insertion of additional genetic material into a normal genome to enhance a trait perceived as desirable. A popular novel, *Geek Love*, by Katherine Dunn, describes how a carnival family saved their show from bankruptcy by deliberately reproducing "special people" who would work as carnival freaks. Lil, the mother, injected certain drugs, arsenic, and radioisotopes into her pregnant body so as to conceive mutant children able to earn a living in carnivals. One child had limbs like flippers, another was a bald albino hunchbacked dwarf. "What greater gift could you offer your children than an inherent ability to earn a living just by being themselves?"[35] The family, in

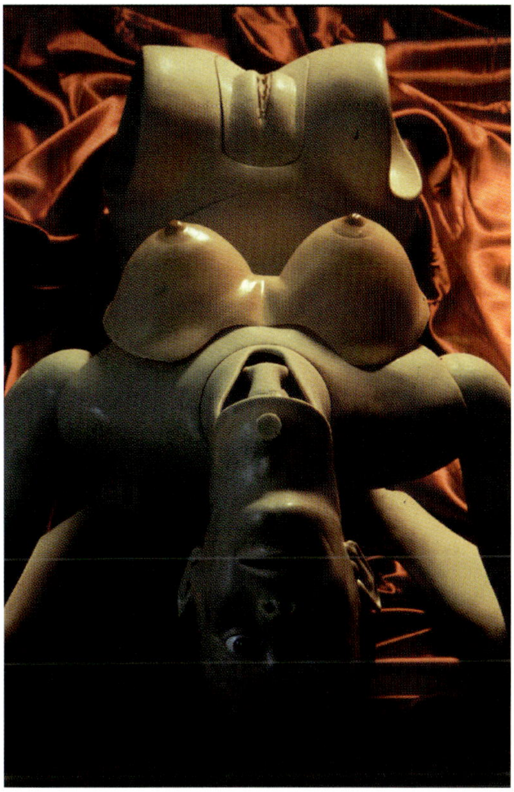

15a Cindy Sherman, *Untitled,* 1992.

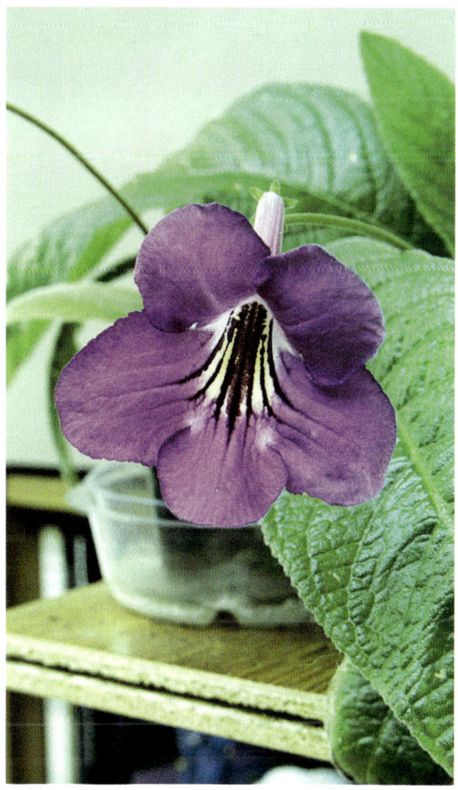

15b George Gessert, *Streptocarpus hybrid 523,* hybridized 1997, first bloomed 1998.

effect, has deliberately created mutants with traits that, in the context of the carnival, are defined as normal.

Artists, too, engage the problematic definition of "the normal." In her "Special Faces" series, Nancy Burson photographs children and adults with congenital cranio-facial disorders. *Untitled (Jessica and Her Mom)* (1991) shows a baby girl with the affliction known as progenia or accelerated aging (Fig. 16). Wearing a lace bonnet, the child is tenderly held up to the camera by an adult who looks directly into her eyes.[36] Burson is addressing the same questions that were raised by Fieldler in his story of "freaks." These questions are also currently being raised by disability groups who insist that "disability is a natural human experience that society should accommodate and not eliminate."[37] What, then, is normality? Is it a situational state, a natural condition, or a healthy ideal? Congenital anomalies, Burson implies, are a normal part of the human condition. In her documentary photographs, she confronts the ambiguities in our definitions of normalcy and pathology, of health, disease, and evolution in the age of genetic engineering.

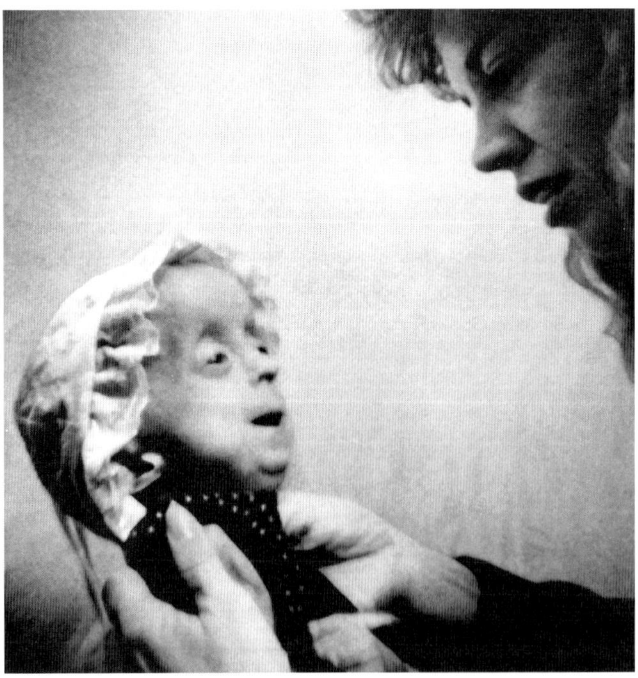

16 Nancy Burson, *Untitled (Jessica and Her Mom)*, 1991.

Burson's photographs of cranial disorders are emphatically different from the sculptured head by Rosemarie Trockel, *Hydrocephalus II* (1982) (Fig. 17). Although it also dramatizes a cranial abnormality, Trockel's sculpture is abstract and ambiguous, with vague facial features that depict abnormality as a solemn, if curious, condition. Molded in plaster and graphite is a swollen head from which issues a blue watery liquid. The artist derived the image both from an African sculpture and from her memory of a hydrocephalic child that she had known when she worked as a nanny. From this experience she became convinced that "the world's sense of normal was totally wrong."[38]

Rosamund Purcell's photographs address diversity, uniqueness, and the arrestingly odd. Interested in the ways in which classification systems are organized by curatorial practices in museums, Purcell explores the ill-defined distinctions between scientific classifications and actual human idiosyncracies. In *Crossing Over: Where Art and Science Meet,* her photographs' "individual distinctiveness" is juxtaposed with text by Stephen Jay Gould.[39] Viewed within the context of the natural world's biological diversity, freaks, they argue, are but a random permutation in nature's gene pool. Her stark photograph of a hydrocephalic child's skeleton (Fig. 18), for the artist, should be viewed in such context.

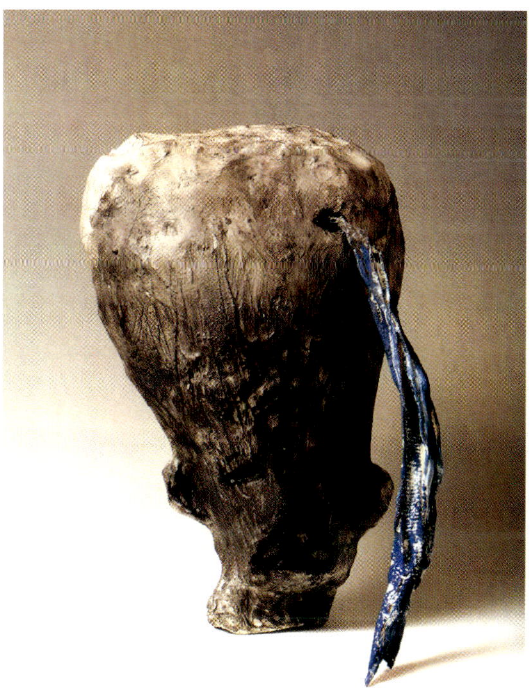

17 Rosemarie Trockel, *Hydrocephalus II,* 1982.

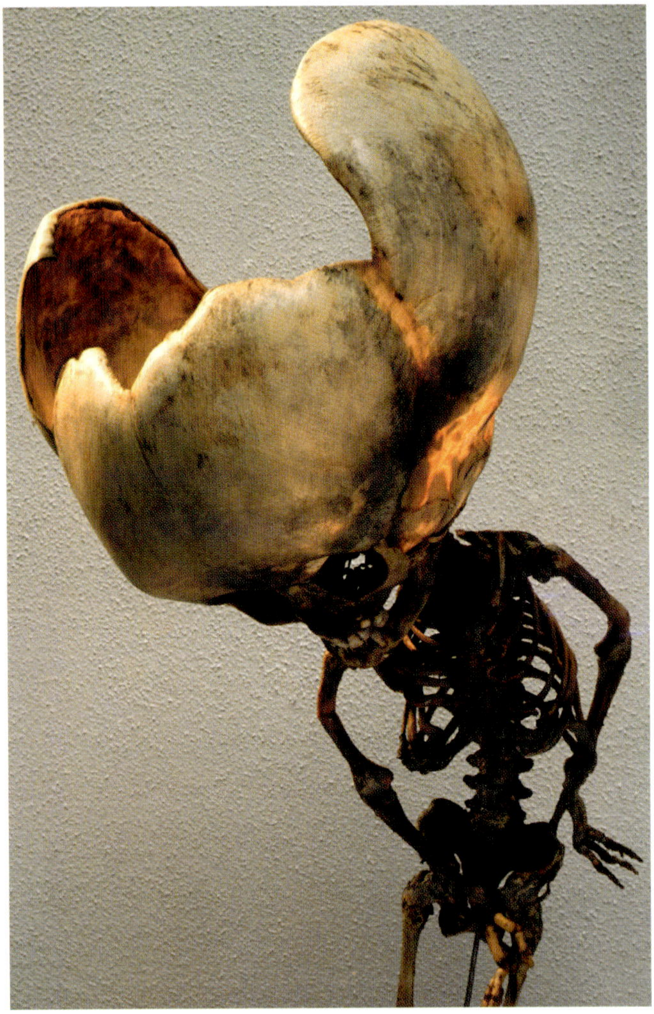

18 Rosamund Purcell, *19th Century Skeleton of Hydrocephalic Child,* 1995.

Joel-Peter Witkin's photographs portray what appear at first glance to be monsters and marvels. The artist mixes motifs emerging from the work of Hieronymus Bosch, Francisco Goya, Diego Velázquez, and Odilon Redon. The images evoke "forbidden territories, realms of the strange and deformed, the sublime and the terrifying" writes art critic Germano Celent.[40] His sources, from the very fringes of society, include people from mental hospitals, sadomasochistic clubs, circuses, and freak shows. However, for the artist these images carry messages of redemption, in that we are all players in life's uncertain and tangled scenarios.[41] In *Portrait of a Dwarf* (1987) (Fig. 19), a female dwarf is dressed in white lace lingerie. Placed between an animal carcass on the left and an idealized but faceless Greco-Roman bust on the right, the female character is situated within a scale of evolutionary ascent. A pair of disembodied mannequin legs completes the tableau, recalling Bellmer's dramatic and distorted depictions of the mechanization and standardization of modern "normal" bodies.

Social Implications of Biotechnology Applications

"Designer babies" are no longer just the stuff of science fiction. The possibilities of genetic manipulation have brought a growing interest in directing the quality of offspring—in acquiring desirable characteristics and even presumed perfection. Anticipated is an ability to endow children not only with health, but also with genes for temperament and perhaps intelligence, personality, or other desired traits. Mary Shelley's Frankenstein monster is no longer a myth, for we have available the resources and the science-based methods to engineer the human body for cosmetic as well as therapeutic purposes.

James Watson, Nobelist and former director of the U.S. Human Genome Project, has dismissed objections to genetic engineering as "utter silliness." "If we could make better human beings by knowing how to add genes, why shouldn't we do it?"[42] "Enhancement is going to happen and nobody can stop it," writes W. French Anderson, one of the first researchers to treat disease with genetic-engineering techniques.[43]

Indeed, gene enhancement has wide appeal in a society where people look to pharmaceutical interventions to improve athletic, cognitive, or sexual performance, and to cosmetic surgery to enhance appearance.[44] Its appeal reflects a convergence of scientific hubris, media hype, and consumer demands. They are encouraged by "Imagologues"—to employ a term by novelist Milan Kundera—those agents of popular culture in the media, the "architects of ideology, building and shaping social visions."[45]

The gene business—the biotechnology and pharmaceutical firms—thrives on cultural beliefs in endless progress, disdain for limits, and demands for technological and unprecedented genetic fixes. The body can be manipulated, reconfigured, and redesigned. Molecular biologist Lee

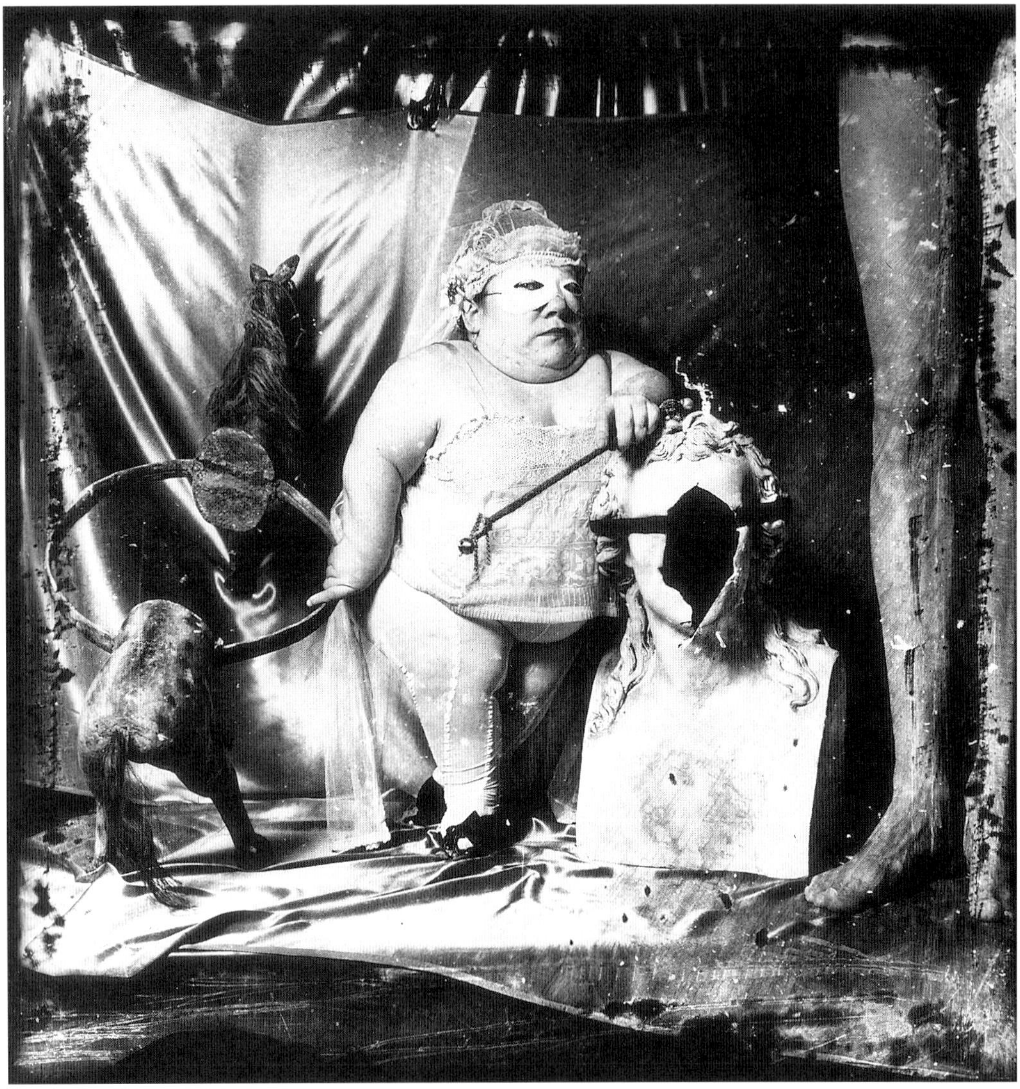

19 Joel-Peter Witkin, *Portrait of a Dwarf*, 1987.

Silver optimistically claims: "We now have the power to seize control of our evolutionary destiny."[46] But the long-term implications of manipulating nature and the hubris of scientists "playing God" have worried philosophers, theologians, and social critics. Should heredity be overcome?

Orlan, a French performance artist, addresses these issues as she manipulates her own body through repeated plastic surgery that she has televised in art galleries in real time. "My work and its ideas, incarnated in my flesh, interrogate the status of the body in our society and its evolution in future generations via new technologies and upcoming genetic manipulation. My body has become a site of public debate that poses crucial questions for our time."[47] In museums such as the Centre Pompidou in Paris and in art galleries throughout the world, Orlan exhibits photographs of her various reconstructed facial features and videos of actual surgical procedures performed on her body. In *Orlan with Imaginary Generic no. 31: Successful Operation[s]* (1990) (Fig. 20), the artist employs images gleaned from art history's canon of masterpieces to fashion her perspective visage. Invoking the forehead of DaVinci's *Mona Lisa*, the chin of Botticelli's *The Birth of Venus*, and the eyes of Gerome's *Psyche*, among others, the artist creates a living art history. No longer confined to painting, sculpture, and photography, ideal types can be reconfigured in the flesh through body enhancement. She explains these painful procedures as her way of questioning the limits of the body's boundaries in the context of the growing possibilities for genetic manipulation. Biology, for Orlan, is no longer destiny.

Rather than tampering with his own body, Michael Rees' notion of sculpture is radically removed from flesh and blood. He constructs his skeletal structures from computer codes, transforming them into physically concrete objects. Unencumbered by the limitations inherent in traditional modeling techniques, Rees directs his computer program to build a sculpture by layering minute layers of material, one on top of the other. In his *Ajna Spine Series 1.11* (1998), the artist instructs the program to grow ears out of a spine (Fig. 21). Like a proprietor in a body-part factory, Rees generates spare parts at will. Mutations, like those of the teratologists, are effortlessly accomplished through computer design.

If the promising implications of genetic engineering have been integrated into futuristic fantasies and cultural expectations, so too have its problematic ethical implications. These could include a decline of tolerance for those who are different, and a growing inequality between those who can afford genetic services and those who cannot.[48] They could violate the ethical obligation that phsyicians and scientists should do no harm. Could it lead to the creation of monsters and mutants? The persistent "Frankenstein" fear of experiments that tamper with nature—with the human body—is a recurrent theme in the public discourse and in the visual arts.

Charles Ray's work can be interpreted in this context. In *Family Romance* (1993) (Fig. 22), he depicts four naturalistic figures. But what is wrong with

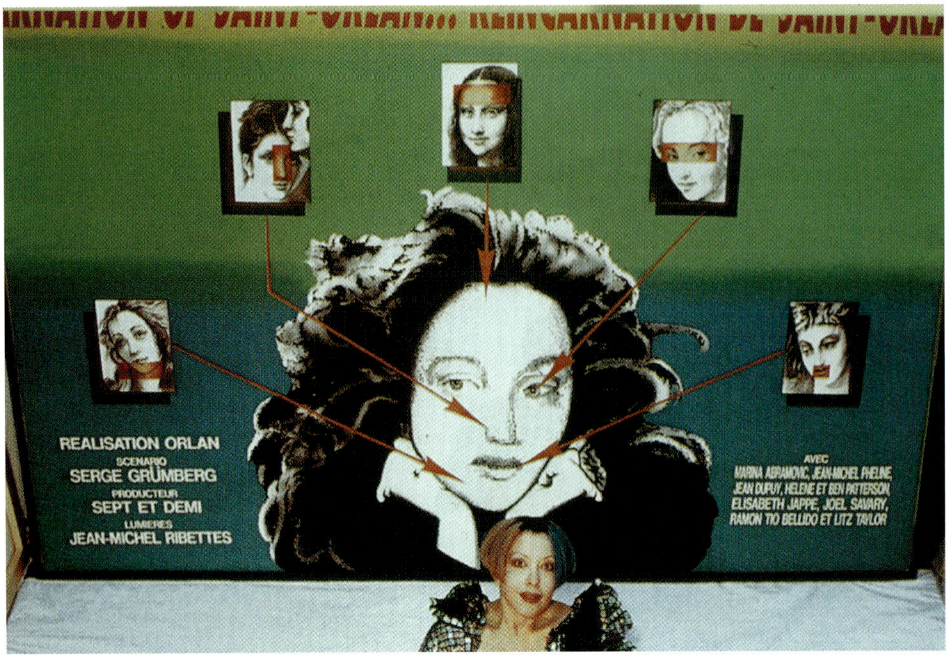

20 Orlan, *Orlan with Imaginary Generic no. 31: Successful Operation(s),* 1990.

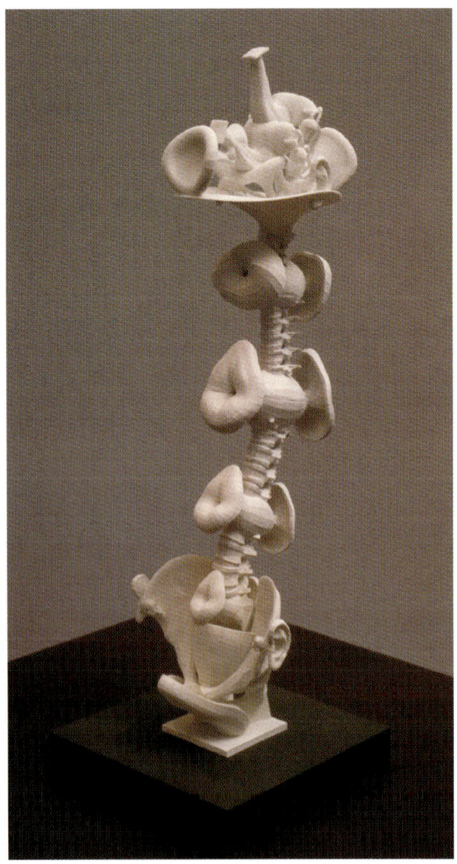

21 Michael Rees, *Ajna Spine Series 1.11,* 1998.

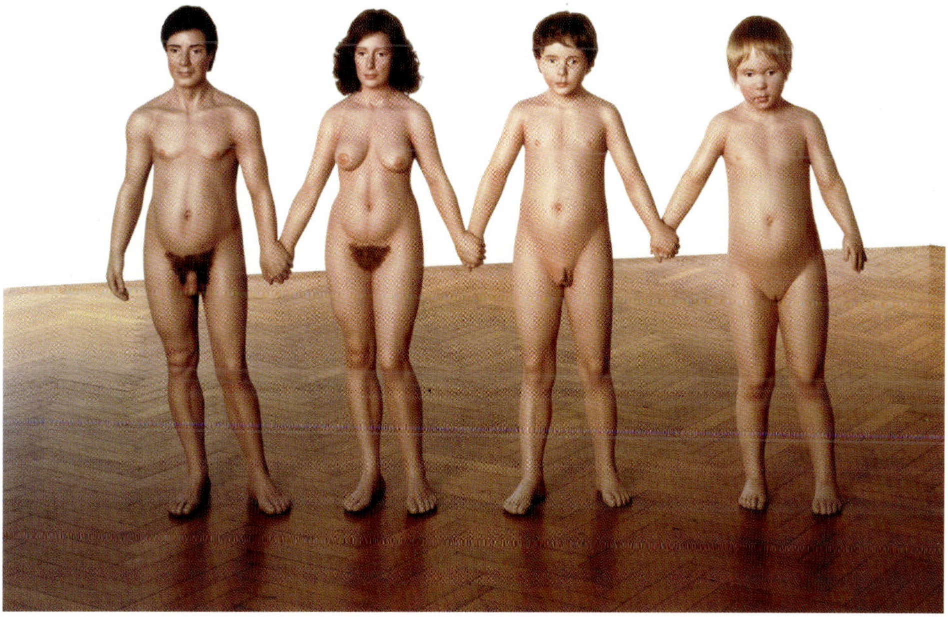

22 Charles Ray, *Family Romance,* 1993.

this picture? By manipulating scale, the young children are proportioned to be the same height as their parents. The figures recall Magritte's 1937 *Spirit of Geometry* (see Fig. 11) in which the artist alternatively transposes the body parts of mother to child. But Ray's sculptured bodies are of supertots. Except for their size, all of their bodily characteristics such as hair, breasts, and baby fat are chronologically normal. The viewer is confronted with the possibility that if these children are to reach adulthood, they would be giants—in effect, monsters in a side show. Have these children been engineered with growth hormones to turn them into the desired offspring? Does a dysfunctional society create a dysfunctional family?

The Mapping of Monsters in the Midway of Science

Research in genetics and, especially, the possibilities of genetic manipulation, have resurrected the long-standing interest in monsters and mutants in popular culture and visual art.[49] Monster fantasies are re-emerging in what Mark Dery calls our "pathological public sphere." "The Great Midway is where we live these days... Barnum's Odditorium lives on."[50] In this age of extremes, the grotesque has reemerged as an object of fascination, a realm for fantasy, and a source of social and ethical anxiety. The Chapman brothers, Joel-Peter Witkin, Charles Ray, Cindy Sherman, and many other artists are reflecting this cultural trend when they represent anatomical anomalies and monsters in their work.

Artists, writes British curator Denna Jones, are "markers of the Zeitgeist—our fears, our hopes, our expectations." They may be sensationalizing science in the settings, not of circuses and side shows, but in galleries of art. As Jones observes: "Science as well as art may at times be guilty of the sensational."[51] The monsters and mutations of contemporary culture and visual art are expressing the ethical dilemmas and the ambiguities that are inherent in science and technology—activities that can cure or kill, create or destroy, provide benefits or cause harm. They are questioning what is natural or constructed, what is normal or perverse. Through visual images—images that are often shocking and offensive—they are commenting on a technology that seems out of control. Should we be tampering with genes? And if we do, will we create "a gruesome parade of horribles" as Chief Justice Burger warned?

Notes

1. Leslie Fiedler. *Freaks: Myths and images of the secret self.* Simon and Schuster, New York, 1978. Anchor Books, 1993.
2. Ibid., p. 347.
3. The Oxford English Dictionary defines a monster as "something extraordinary or unnatural, a prodigy or a marvel." Oxford University Press, London, 1970.
4. Lorraine Daston and Katherine Park. *Wonders and the order of nature.* Zone Books, New York, 2001.

5. George Annas. 1990. Of monster mythology. *Emory Law Journal* 39: 629–665.
6. Jan Bondeson. *A cabinet of medical curiosities: A compendium of the odd, the bizarre, and the unexpected.* W.W. Norton, New York, 1997, p. 200.
7. These debates over the theory of evolution were about the relationship between the evolution of species and the life history and development of individual organisms.
8. "Foreign Body," curated by Carol Squiers at the International Center for Photography, New York, January 11–March 16, 2002, documented the continued public fascination with unique body characteristics. The images were drawn from medical photography and included people with facial and body anomalies, extreme height, and congenital disease.
9. Geoffroy Saint-Hilaire. *Histoire générale et particuliére des anomalies de l'organisation chez l'homme et les animaux: Ouvrage comprenant des recherches sur les caractéres, la classification, l'influence physiologique et pathologique, les rapports généraux, les lois et les causes des monstruosités, variétés et vices de conformation, ou traité de tératologie.* Bruxelles: Establisement encyclographique, 1837–1838.
10. Fiedler, op. cit. Chapter 9, "From theology to teratology," pp. 229–255.
11. Dorothy Nelkin. 1985. Popular images of the atom. In *Social responses to technological change* (ed. A. Brannigan and S. Goldenberg). Greenwood Press, Westport; also Spencer Weart. *Nuclear fear: A history of images.* Harvard University Press, Cambridge, 1988.
12. Cornelia Hesse-Honegger. *Heteroptera: The beautiful and the other, or images of a mutating world.* Scalo, Zurich, 2001.
13. Jon Turner. *Frankenstein's footsteps: Science, genetics and popular culture.* Yale University Press, New Haven, 1998, p. 128.
14. Theodore Roszak. 1974. The monster and the titan. *Daedalus* Summer 1974, p. 31.
15. Lawrence Weschler. *Mr. Wilson's cabinet of wonder.* Vintage Books, New York, 1995. Also see Stanley B. Burns. *A morning's work: Medical photographs from the Burns archive and collection,* 1843–1939. Twin Palms Publisher, Santa Fe, 1998.
16. Cary Keirns. Babies in bottles and sharp metal objects. University of Pennsylvania manuscript, 1997.
17. Gretchen Worden. *The Mutter Museum of the College of Physicians in Philadelphia.* Blast Books, New York, 2002. Gretchen Worden is Director of the Mutter Museum. She has invited a number of artists such as William Wegman, Arne Svenson, Mark Kessell, Gwen Atkin, and Allan Lutwig, among others, to photograph the museum's collections.
18. Lori Andrews and Dorothy Nelkin. *Body bazaar: The market for human tissue in the biotechnology age.* Crown Books, New York, 2001, Chapter 7.
19. See the work of 16th century artists Hieronymus Bosch and Pieter Brugel the Elder. James Luther Adams and Wilson Yates. *The grotesque in art and literature: Theological reflections.* William B. Eerdman's Publishing, Grand Rapids, Michigan/Cambridge, U.K., 1997, pp. xiii–xv.
20. Ibid. A forthcoming symposium, cohosted by the School of Visual Arts and the Museum of Modern Art will take place in Feb. 2004. Panelists to date are film critic Amy Taubin, curator Rob Storr, author Mary Roach, and artist/writer Suzanne Anker.
21. Robert Dory. *Human concern/Personal torment:The grotesque in American art.* Whitney Museum of American Art, New York, 1969. Also see Nina-Athanassoglou-Kallmyer's chapter on "Ugliness" in Robert S. Nelkson and Richard Shiff, editors, *Critical terms for art history* (2nd edition), University of Chicago Press, Chicago, 2003. pp. 281–295.

22. Douglas W. Druick and Peter Kort Zegers. 1994. In the public eye. In *Odilon Redon: Prince of dreams 1840–1916* (ed. D. Druick). Abrams and the Art Institute of Chicago, Illinois, pp. 137–140.
23. Maud Lavin. *Cut with a kitchen knife: The Weimar photomontages of Hannah Hoch.* Yale University Press, New Haven, 1993, pp. 165–166.
24. Psychoanalyst Karl Figlio defined the "uncanny" as "a special kind of anxiety related to feeling stalked by something familiar. Karl Figlio. 1996. Knowing, loving and hating nature: A psychoanalytic view. In *Future natural: Nature, science and culture,* (ed. G. Robertson et al.) Routledge, New York, 1996, pp. 78–80.
25. Hans Bellmer quoted in Therese Lichtenstein. *Behind closed doors: The art of Hans Bellmer*, University of California Press, Los Angeles, 2001. Other interpretations of these disturbing and controversial images suggest that they represent misogyny, commodification, and alienation as the person is equated with a mannequin.
26. Douglas Dreishpoon. Organic inflections. In *Sculpture: Inside and outside.* Walker Art Center, Minneapolis. 1988. Also see Jennifer Mundy. Form and creation. The impact of the biological sciences on modern art. In *Creation: Modern art and nature.* Exhibition catalogue, National Gallery of Modern Art, Edinburgh, 1984.
27. Historian Lawrence Alloway describes this term as follows; "Bio: a combining form denoting relation to, or connection with, life, vital phenomenon, or living organisms. Morphology: the features, collectively comprised in the form and structure of an organism or any of its parts." See "The Biomorphic Forties," *Artforum* December, 1967.
28. Jeffrey Weiss. 1983. Science and primitivism: A fearful symmetry in the early New York school. *Arts Magazine* 57: 86.
29. Diamond v. Chakrabarty, 448 U.S. 303, 309-10 (1980).
30. Dorothy Nelkin and M. Susan Lindee. *The DNA mystique: The gene as cultural icon.* W.H. Freeman, New York, 1995.
31. Rona Pondick. *Works, 1986–2001.* Sonnabend Press, New York, 2001.
32. Review in *Sculpture Magazine.* 17: 76. Also see Saatchi Collection, Normal Rosenthal and Brooks Adams. Royal Academy of Arts Great Britain, Richard Shone. *Sensation: Young British artists from the Saatchi collection.* Thames & Hudson, London, 1998.
33. David Freedberg. *The power of images: Studies in the history and theory of response.* University of Chicago Press, Illinois, 1999.
34. George Gessert. 1993. Notes on genetic art. *Leonardo Mag.* 26: 205–211. http://mitpress2.mit.edu/e-journals/Leonardo/index.html
35. Katherine Dunn. *Geek love.* Warner Books, New York, 1983. Reprinted by Vintage Press, 2002.
36. Kirby Gookin. 2002. Review. *Art Forum,* May, 2002, p. 183.
37. Adrienne Asch. 1995. Disability: Attitudes and sociological perspectives. In Warren Thomas Reich. *Encyclopedia of bioethics 2,* revised edition. Macmillian, New York, p. 605.
38. Quoted in Sidra Stitch, ed. *Rosemarie Trockel.* Prestal-Verlag, Munich, 1991, p. 25.
39. Rosamund Wolff Purcell and Stephen Jay Gould. *Finders, keepers: Treasures and oddities of natural history.* W.W. Norton, New York, 1992, p. 139. Also Stephen Jay Gould and Rosamund Wolff Purcell. *Crossing over: Where art and sciences meet.* Three Rivers Press, New York, 2000, pp. 24–29.
40. Germano Celent. *Witkin* Scalo Edition, Zurich-Berlin-New York, 1995.
41. Conversation between Suzanne Anker and Joel-Peter Witkin, July, 2003.
42. James Watson, quoted in *Engineering the human genome,* Summary Report, UCLA, June 1999.

43. W. French Anderson, cited in Vincent Kiernan. Cosmetic uses of genetic engineering may soon be a reality. *Chronicle of Higher Education*, October, 3, 1997, pp. A17–18.
44. Sander Gilman. *Making the body beautiful.* Princeton University Press, New Jersey, 1999.
45. Milan Kundera, in his novel *Immortality*, invented the word "imageology" to capture this phenomenon. "Imagologues create systems of ideals and anti-ideals... that influence our behavior, our political opinions and aesthetic tastes."
46. Lee Silver. *Remaking Eden: How genetic engineering and cloning will remake the American family.* Avon Books, New York, 1998.
47. Orlan, interviewed in Peg Phelan and Jill Lane, eds. *The ends of performance*, New York University Press, New York, 1998, pp. 319.
48. Maxwell Mehlman and J. Botkin. *Access to the genome: The challenge to equality.* Georgetown University Press, Washington, D.C., 1998.
49. Also see the work of Dingh Q Le and William Latham with regard to mutation.
50. Mark Dery. *The pyrotechnic insanitarium.* Grove Press, New York, 1999, p. 91.
51. Denna Jones. Working drafts: Envisioning the human genome. The Wellcome Trust, Catalog to the Two-Ten Gallery exhibition, September to January 2002.

Blurring Boundaries: Chimeras and Transgenics

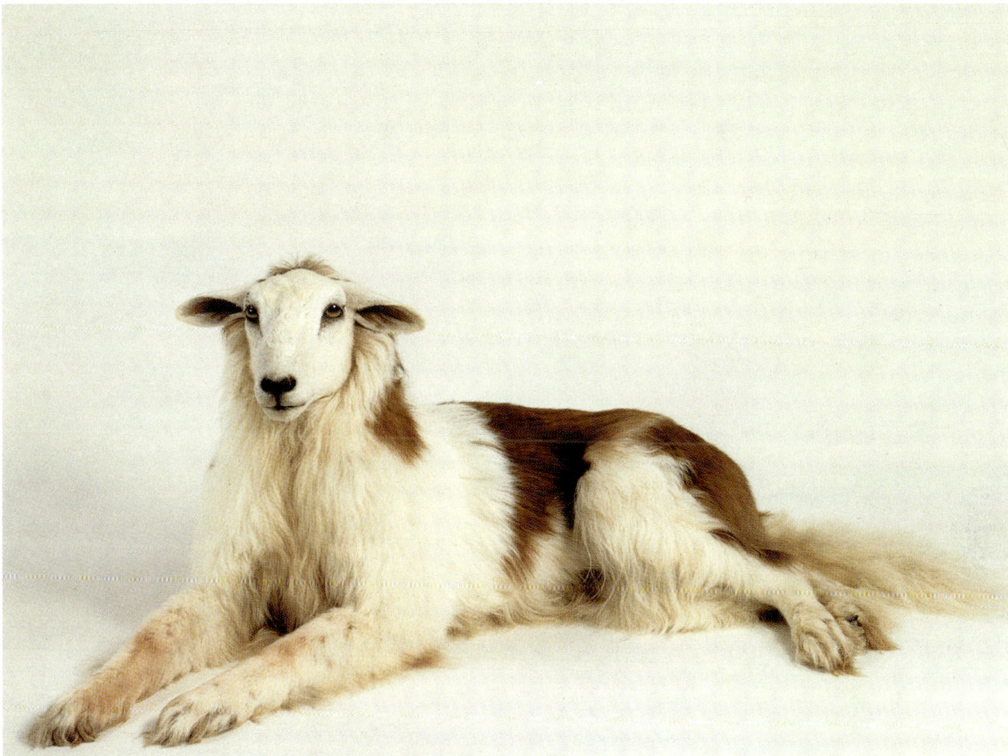

1 Thomas Grünfeld, *Misfit* (*St. Bernard*), 1994.

THOMAS GRÜNFELD'S *Misfit* (*St. Bernard*) (1994) is an odd animal with the body of a large Saint Bernard dog and a sheep's slender head. (Fig. 1). The sculpture is ambiguous and incongruous, pieced together from taxidermied animal parts. Genetically, the Saint Bernard was bred to maintain a thick heavy head with a short muzzle. Popular stories convey an image of the massive animal, wearing a keg of whiskey around his collar, coming to rescue frozen and hapless Alpine climbers. But Grünfeld's *Misfit* is a graceful animal with its narrow refined head, horizontally erect ears, and almond-shaped eyes. *Misfit* is an elegant but unnatural animal. *Misfit* is a classic chimera.

From Mythology to the Laboratory

The Chimera is a fire-breathing she-monster from Greek mythology, often represented as seamlessly comprising the body of a goat, the head of a lion, and the tail of a serpent.[1] Born from the union between the giant Typhaon and the half-serpent Ekidna, both hybrids themselves, this imaginary beast had the capacity to transcend and ignore species boundaries, and so embodied dangerous spiritual and emotional qualities. As the namesake of a long line of these mixed-species creatures, the word "chimera" has come to mean a symbolic monster composed of incongruous parts. Other examples include the Centaur, the Griffon, the Sphinx, and the Minotaur. From the ancient world through postmodern culture, chimeras have been and continue to be magical and monstrous figures in the visual arts.

Today the chimera has moved from the envisioned world of mythology to the real space of scientific laboratories, from its representation in legend to its role in biological research. In 1907, the term chimera referred to hybrids—plants created through grafting together two separate vegetative varieties. By 1968, chimera became a term in genetics, used to describe organisms that were made up of cells and tissues from two or more species. To create genetically mixed organisms, research scientists manipulate life forms by conjoining cells from genetically distinct embryos, usually at the blastocyst stage. For example, mice chimeras are fabricated in the lab by introducing cells from other strains or species into their cell lines.

So-called "chimeras" were once categorical freaks—among monsters in 19th-century side shows and other public spectacles. Man-monkeys, a dog-faced boy, a lion-faced man—all were paraded and presented as human/animal hybrids, although this was not their genetic condition. They were the most popular of the freaks, but they differed from the mutations described in the previous chapter both biologically and in their representation in art.[2] As "species that straddle the line between us and our animal brothers,"[3] they transgressed "natural" boundaries. Such creatures were especially fascinating subjects for scrutiny because they seemed to yield insight into processes through which one animal species evolves into another. The famous Wild Men of Borneo, for example, were believed to be "Darwin's missing link." Nineteenth-century biologists and physicians assumed, incorrectly, that people with physical mutations which left them with vestigial organs such as tails or horns were sporting relics of their primate ancestors. Such people, crossing classification boundaries, were specimens, and their complex mutations were regarded as means to unravel, while revealing, evolution's secrets.[4]

Also intriguing to biologists (as well as voyeurs) were people ambulating with ambiguous sexual, racial, or animal-like features. From the early Greeks to the Romans until the late 19th century, medical scientists observed such people, first trying to explain their position in nature and later trying to actually fabricate them in the laboratory. Was nature just brutal and capricious, or could these anomalies be logically explained? For centuries, common theories attributed cross-species abnormalities to "maternal impressions."[5] If a pregnant woman experienced a frightening image of a terrifying animal, for example, this image would be imprinted in her fetus. Such a baby would then develop related characteristics, in effect manifesting the characteristics of two separate species. This, at the time, seemed to explain the issue of babies with cat-like heads, extra fingers, porcupine skin, and various skin diseases. The theory was also used to account for skeletal distortions of the so-called Elephant Man, Joseph Merrick; his mother was nearly trampled by an elephant while she was pregnant.[6]

These fabled chimeras were vehicles for bringing revenue to America's early and fledgling museums. P.T. Barnum, the entrepreneur whose name became synonymous with the circus, developed popular entertainment programs to attract paying visitors. Barnum's American Museum (1841–1865), located in New York City, was a mixture of natural history and spectacle. By displaying the infamous "Feejee Mermaid" (Fig. 2), an alleged cross between a woman and a fish, Barnum tripled revenues at his museum—especially after scientists attacked the exhibit as a fraud.[7] The Feejee Mermaid is still featured in a "museum of the unusual" called the "Freakatorium" in New York City, along with other artifacts and ephemera including Siamese twin babies in a jar, a two-headed cow, and a shrunken human skull.

Coincidentally, German artist Rosemarie Trockel's *Untitled (Pennsylvania Station)* (1987) (Fig. 3) recalls Barnum's primitive figure. Her sculpture is a tortured mutation that blurs the distinction between woman and fish. Although it resembles the Feejee Mermaid, Trockel's grotesque sculpture is a social commentary and not intended as a hoax. With her skeletal head and burned facade, this violent, unsettling image is a brutal attack on society's instrumental use of a woman's body and a darkening comment on those primitive forces that seem to pervade modern history.

The very idea of human/animal hybrid species has evoked a certain horror. In 1896, H.G. Wells' evolutionary parable, *The Island of Dr. Moreau*, described the "beast men" of his fantasy island:

> The disproportion between the legs of these creatures and the length of their bodies...the forward carriage of the head and the clumsy and inhuman curvature of the spine...the deformity in their faces almost all of which were prognathous, malformed about the ears, with large and protuberant noses.... each preserved the quality of its particular species: the human mark distorted but did not hide the leopard, the ox, or the sow.[8]

H.G. Wells wrote this description (strikingly reminiscent of the animal-like human portraits by 17th-century artist Charles le Brun) as a warning about the implications of experiments that blur the distinctions between

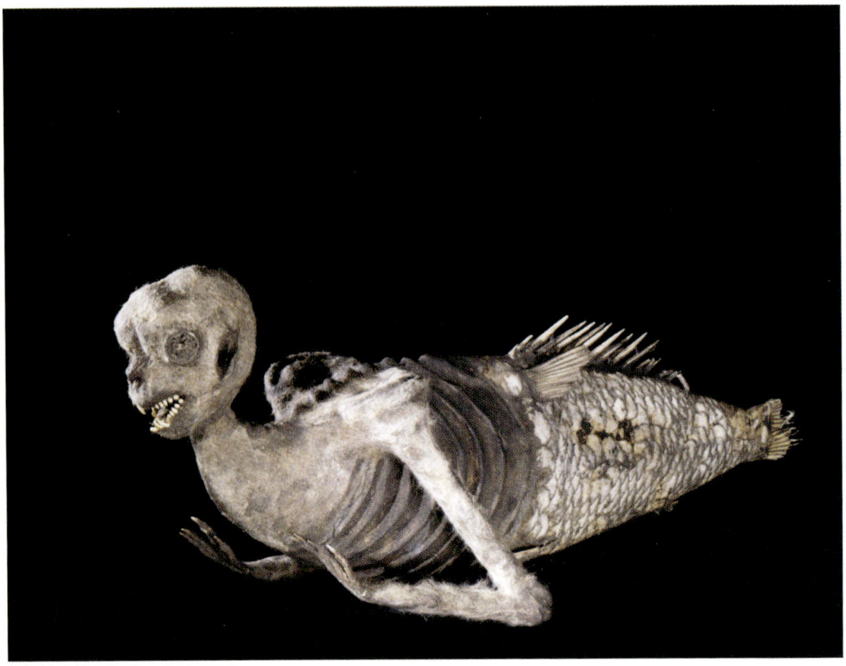

2 Artist unknown, Feejee mermaid.

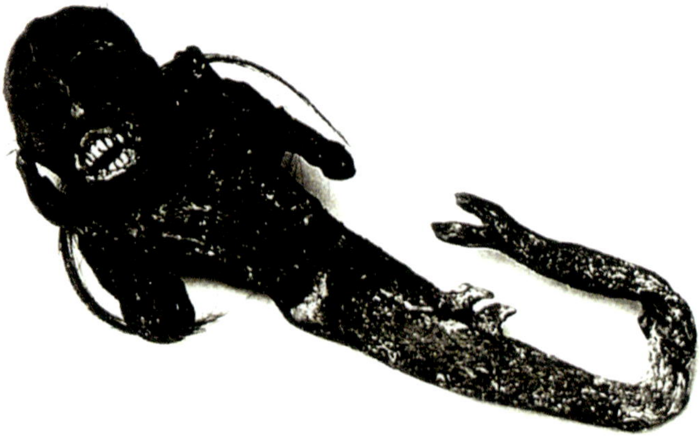

3 Rosemarie Trockel, *Untitled (Pennsylvania Station)*, 1987.

man and beast. le Brun's drawings were imaginary projections of phrenological theory that associated human emotions with animal types (see earlier chapter on reductionism). But Dr. Moreau's endeavor was different. He arrived at his fantasy island expressly to develop techniques for creating novel human beings from animal creatures. The "beast people" were the ill-fated result of such efforts.[9] In Wells' science fiction fantasy, the chimera is a powerful warning about the grim and drastic consequences of scientific manipulations.

Early examples of such manipulations could be called figurative hybrids, involving mechanical replacements of human parts with animal tissue. During the 1920s, medical scientists witnessed therapeutic possibilities in such animal-to-human transplantation, engaging in a variety of experiments employing animal organs and their tissues to replace diseased tissues in human beings. For example, they grafted monkey glands onto diseased human organs in the hope of restoring normal function. Chimps and pigs were also used, with tragic consequences, as sources of blood for human transfusions, while surgeons transplanted animal kidneys into dying patients. Despite many disasters, such procedures were a source of hope for terminally ill patients. Today, called xenotransplantation, similar, but more thoroughly understood procedures are generally welcomed with optimistic expectations.

Early experiments in the clinical use of animal parts were also fodder for political satire. Russian novelist Mikhail Bulgakov's *Heart of a Dog* (1925) was a tale about a Russian scientist who was famous (or infamous) for rejuvenation treatments. He treated women with sexual disorders by transplanting monkey ovaries. One day, he picked up a stray dog, Sharik, and brought him to his lab as a subject for experimentation. He replaced Sharik's pituitary gland and testes with those of a criminal. The dog became Polygraph Polygraphovitch Sharikov, assuming both the persona and the physical characteristics of a vulgar, lecherous human being: "a man with the heart of a dog." Sharik, as a chimera, was a bitter satire of the malefic excesses and subsequent hubris of scientific inquiry.[10]

From Collage to Chimera: Cut and Paste

As an archetype of an illusion, a fabrication of the mind, a fantastic fabled hallucination, and increasingly an experimental reality, the chimera has long been a source of fascination in science, literature, and the visual arts. Reformatting fragments into alternative narratives in the arts is a technique familiarly known as collage. Early modernist artists employed these diverse methods of recontextualization and alteration, including photocollage, montage, the objet trouvé, and the ready-made. Fascinated by the versatility of exchangeable parts, they invented images of a host of organisms consisting of unique aggregations of existing materials.[11] It is interesting to note that the cut-and-paste character of transgenic experiments mirrors avant-garde practices in art during the 1920s. These, after all, were the early days of the assembly line and mass production techniques based on the interchangeability of product components.

Many artists from the Dada and surrealist movements[12] reconstructed the body as a combination of disparate elements. In these confabulous images, human beings meld with machines as robotic extensions, or converge with animals as mythical beasts. The fantastic, the marvelous, come into being in the most unlikely ways to address the discontinuities, ambiguities, and anxieties associated with scientific experiments, technological developments, and the oscillating changes brought about by the industrial revolution.[13]

For example, Max Ernst, in *Stratified Rocks, Nature's Gift of Gneiss Lava* (1920) (Fig. 4), depicts extravagant and preposterous organisms by intermingling images of specimens from the animal, vegetable, and mineral kingdoms. Body parts such as rib cages and blood vessels are juxtaposed with fabricated exotic plants and sedimentary rocks—all superimposed on an existing anatomical engraving. Clear-cut objects give way to enigmatic biomorphic, skeletal, and geological forms. In this mixed media palimpsest, Ernst creates an "eery world of fantastic beasts."[14]

In his study of Ernst's techniques, art historian Werner Spies defines the principle of collage in terms that could be used to describe transgenic experiments: Two distinct and unrelated forms are made to meet, displacing their original meanings. The coupling results in a new entity evoking disparate information requiring alternate interpretations.[15] Ernst, in effect, converts collage into a semantic and syntactic collision. His fragments and dislocations generate chimeras—some lyrical, some perverse. For example, in *Here Everything is Still Floating* (1920) (Fig. 5), Ernst turns a beetle on its back and transforms the living insect into a technologically produced steamship. A nearby fish floats by with an exposed skeletal spine, an image evocative of both X-rays and Australian aborginal "dream-time" paintings: Inside is outside, up is down. Ernst's fantastic beasts, fabrications of his imagination, disrupt our expectations and expose the multiple identity of visual images.

Similarly, Hannah Höch uncovers secretive meanings through the structural principle of collage. She creates a dialectic of conflicting information in her *Monument I* (1924) (Fig. 6) by presenting a body that is an incongruous amalgam of fragmented parts. The head is a tribal mask; the torso is female, but it has one female leg, one partial male leg in folded trousers, and one hoofed animal foot. Positioned on a pedestal—that is, as an exhibit on display—a perplexing figure provokes bewilderment and uncertainty. By disrupting the integrity and unity of a singular creature, Höch's disturbing chimera is an ambiguous and formidable being, open to perpetually fluctuating interpretations.

Raoul Hausmann's *Tatlin at Home* (1920) (Fig. 7) is a collage constructed from mass media-produced images of war machines, strategic maps, and

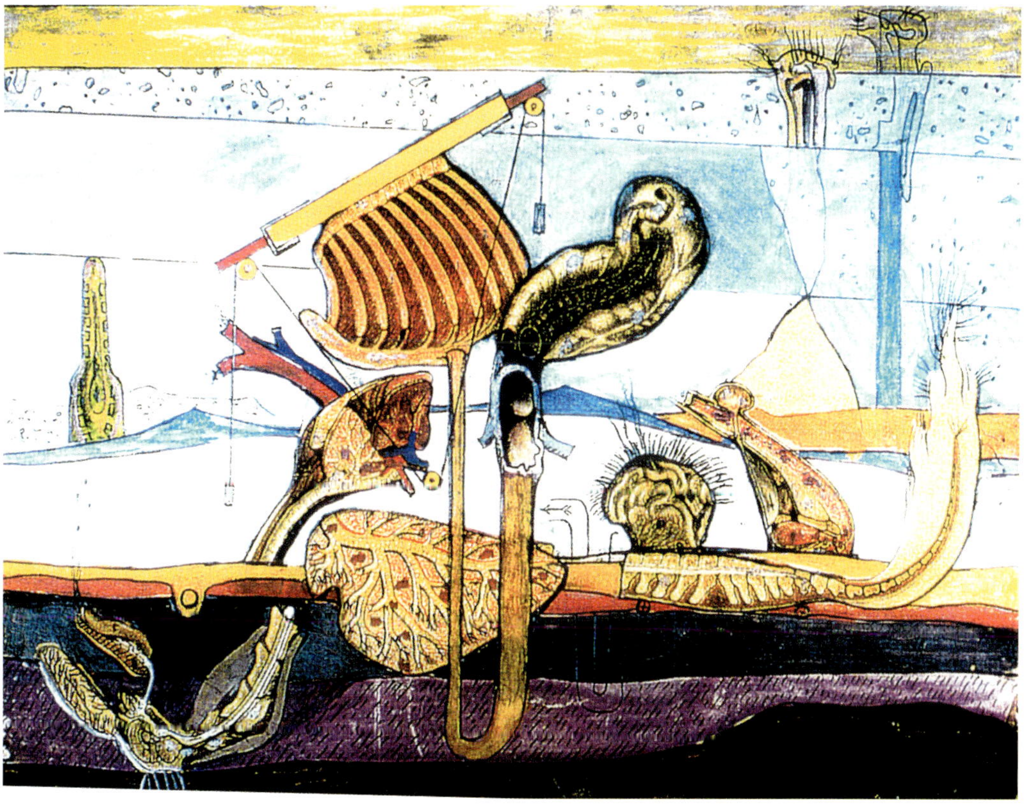

4 Max Ernst, *Stratified Rocks, Nature's Gift of Gneiss Lava,* 1920.

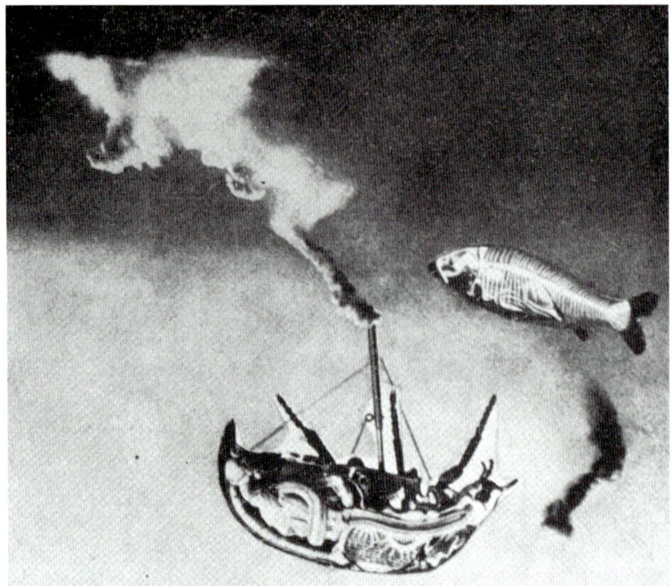

5 Max Ernst, *Here Everything is Still Floating (Le Vapeur et le Poisson),* 1920.

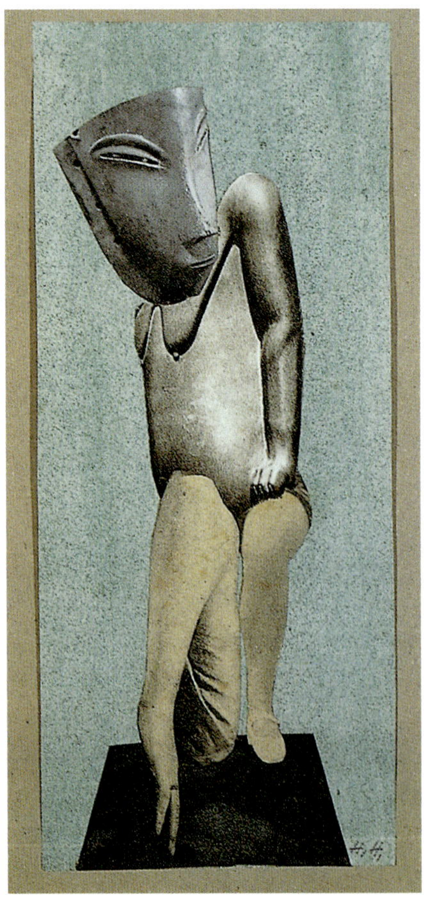

6 Hannah Höch, *Monument I,* 1924.

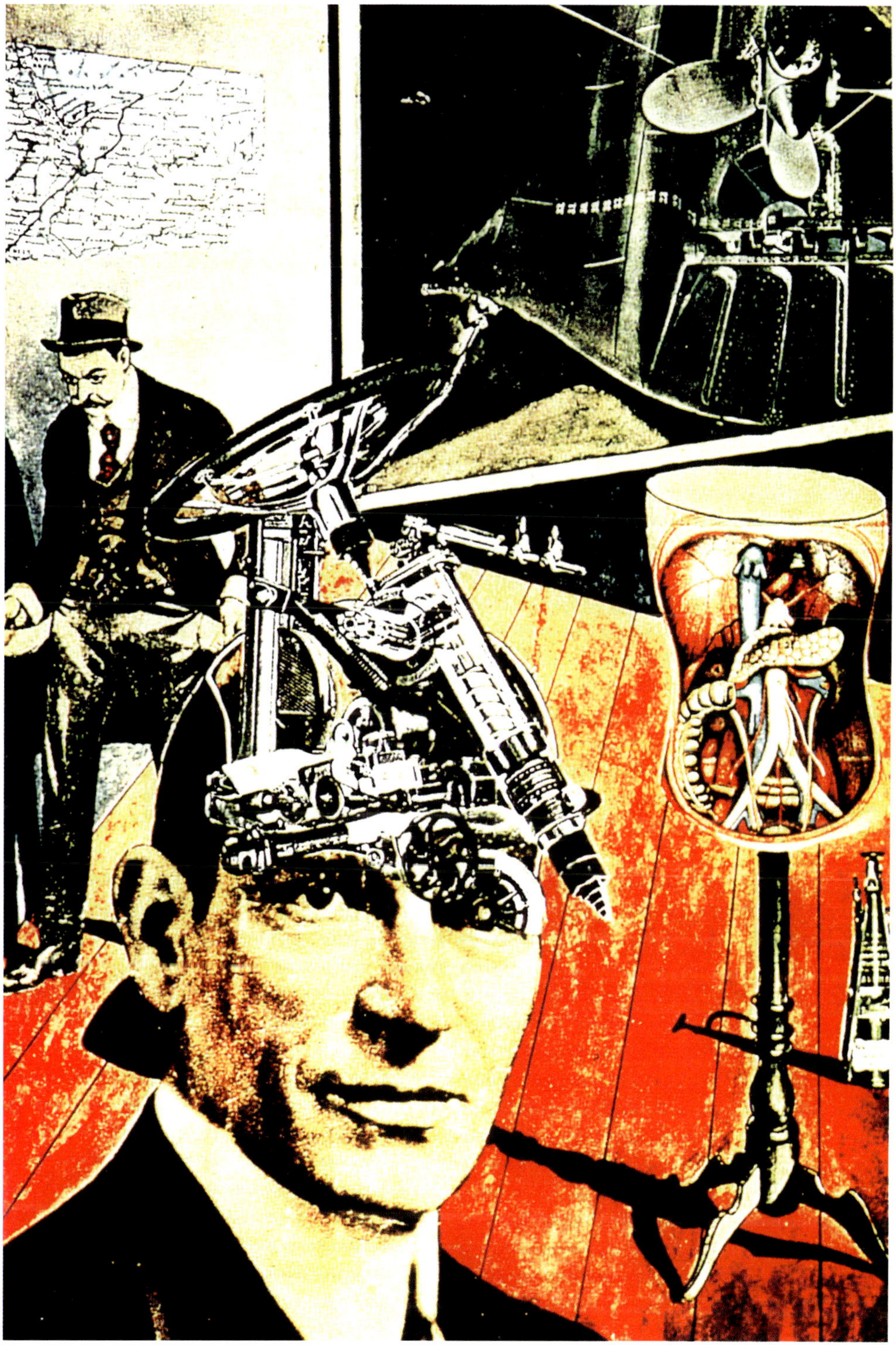

7 Raoul Hausmann, *Tatlin at Home*, 1920

organic viscera. The work prefigures cyborg imagery in its blurring of boundaries between organism and machine. As part of the Berlin Dada group, Haussman was intrigued by the political power of recontextualized mass-media images. Reacting fervently against the war and its resulting perils, Haussman employed images that are routinely generated in bourgeois society's publications. With a twist and a snip of publicly printed matter, he was able to comment on the madness of the postwar world.[16]

Montage, like collage, unites eclectic parts through cut-and-paste techniques. Devised by the Russian filmmaker Sergei Eisenstein in the early 20th century, montage refers to the deconstruction or splicing of film and its reassembly in alternative patterns that violate the temporal frame in which the film was shot. Using these techniques, Eisenstein could create illusions of experience through editing, much as current news broadcasts do when taking the viewer around the world via newsclips. Like the reshuffling of RNA transcription, the film edit recombines fragments to create new narratives.[17]

Employing such revolutionary art-making techniques, visual artists of the 1920s created chimeric images with widely unexpected forms from dislocated parts of heterogeneous elements. Altering, recombining, duplicating, enhancing, recontextualizing, and amplifying were all techniques through which artists sought to extend existing limits of representation. Geneticists today use parallel methods, combining heterogeneous elements to create transgenic creatures.

Transgenic Chimeras

Transpecies creatures are no longer mythical icons or figments of human imagination: They are, as well, the genuine results of transgenic experiments.[18] Biotechnology applications involve the creation, combination, and construction of animals to contain the genetic traits of several species, including human beings. Similar to a collage, transgenic animals are an aggregation of extant materials. Genes are reassembled and recombined to create more efficient and profitable "super-animals" for food or pharmaceutical production, to develop research models for neurodegenerative diseases, and to expand further the sources of tissue and organs rendered available for human transplantation.

Transgenic techniques have to a great extent been employed on the farm. Bioengineers have created sheep–goat chimeras, known colloquially as "geeps," and transgenic pigs that produce low-cholesterol meat. More recently, scientists are implanting human genes into animals so that they become, in effect, biological factories for the production of body substances, therapeutic materials, or pharmaceutical products. Human genes are inserted in cows to increase their production of milk. Mice are created to produce the human blood protein TPA, and cows are altered to produce human hormones.[19] As part of the war on terrorism, cows are also engineered to produce antibodies to botulism, and human genes are added to assure they can be safely used in human vaccinations in the event of terrorist attacks. Among the proposals for stem cell research are experiments that would involve creating human–mouse hybrids—chimeras that would have human cells saturated throughout their bodies. Such hybrids, created by injecting human embryonic cells into mouse embryos, could assist scientists to follow the behaviors of human cells blighted by genetic diseases.[20]

Transgenic animals can also be employed as factories to produce the eggs of endangered species. These could then be fertilized and used to impregnate endangered animals. In a Dr. Seuss-like experiment, scientists coaxed a mouse into growing an elephant egg by transplanting ovarian tissue from African elephants—an endangered species—into laboratory mice that are specially bred so they would not reject foreign tissue. The different scale of the two beasts was no problem, since eggs of both species are microscopic.[21]

To resolve the shortage of human organs available for transplantation, research scientists have once again turned to xenotransplantation, the use of animal organs and tissues in human beings. In the 1960s, surgeons transplanted chimp kidneys into six patients, one of whom survived for a record of nine months. But there has been infrequent subsequent success. A baboon heart, transplanted into Baby Fae in 1984, was rejected in a mere three weeks. Today, scientists are creating transgenic pigs engineered to prevent rejections of their tissue.[22] In 2001, scientists at PPL Therapeutics, an Edinburgh company that helped to create Dolly, the cloned sheep, created cloned piglets. This research was intended to produce pigs that lack genes for encoding the porcine cell-surface antigen, which triggers rejection when pig tissue is implanted in a human body. In light of recent developments in immunology and transgenics, some scientists regard xenotransplantation as "the next great medical revolution."[23]

Despite their therapeutic potential, transgenic experiments are controversial because of numerous concerns about their high failure rate and risks of cross-species infection. Transgenic procedures are also controversial because they seem to contradict the very concept of species, violating assumptions about "natural" breeding barriers. Experiments that involve human genes raise especially difficult questions. If one introduces human elements into a mouse or a pig, could the resultant animal share any attributes with human beings as in Bulgakov's fantasy? If a transgenic animal is ever sold as meat, could this be a form of cannibalism? By scrambling the traditional demarcations of living systems, transgenic experiments threaten species integrity, undermining our concepts of nature and raising disturbing ethical dilemmas.[24]

Small wonder that these procedures have drawn protracted political attention. During 1998 Congressional hearings on the possible need to legislate transgenic experiments, U.S. Representative Charles Rose (D-NC) warned that "We are engineering ourselves away from natural selection into a mechanical selection of traits." Reverend Granberg Michaelson of the

National Council of Churches testified at these hearings: "Cows do not mate with fish. Humans do not mate with pigs. Fireflies do not mate with tobacco plants. These combinations are more than what can be called simply natural occurrences."[25]

Similarly, popular culture's response to transgenics, in the tradition of H.G. Wells, has often been negative—critical of scientific experiments that challenge deeply held convictions about what is "natural," subsequently violating fundamental assumptions about the distinctions between man and animals and interfering with such boundaries encompassing human life. In a 1991 filmed version of his novel, *Chimera*, Stephen Gallagher creates a man–ape hybrid which suggests that "scientists are in a furious race to create a species of powerful sub-humans."[26] Spiderman, comic book hero and the star of the blockbuster 2002 summer film, was a student who was bitten by a bioengineered spider in a research lab. As a result, he assumed some of the attributes of that spider.[27]

Bioluminescence: Or, What Does Green Mean?

Among the most useful techniques employed in transgenic experiments is "bioluminescence." Bioluminescence, sometimes referred to as "cold chemical fire" or "living light," is a diagnostic tool which transfers luminescent jellyfish genes that code for green fluorescent protein (GFP) into research subjects to be used as markers.[28] These phosphorescent markers can then be monitored by magnetic resonance imaging techniques, allowing research scientists to peer inside living cells to discover how diseases develop and evolve (Fig. 8). Bioluminescence has always fascinated observers. The Roman naturalist Pliny the Elder commented on this substance nearly 2000 years ago in reference to a phosphorescent clam. The creature became a culinary novelty in the 1st century as Romans threw clam feasts and sported glowing green mouths.[29] Luminescent click beetles were hair adornments in the Caribbean. During World War II, Japanese soldiers used the residue of luminescent crustaceans to read maps at night.[30] These days, bioluminescent toys that glow in the dark enjoy popular cultural appeal.

In a 1986 laboratory experiment, geneticist David Ow combined tobacco and firefly genes to produce plants that glowed in darkness (Fig. 9).[31] In 2000, research scientists at the Oregon Regional Primate Center created an infant rhesus monkey that carried GFP. Named ANDi (a backward acronym for inserted DNA), this bioluminescent monkey, so its creators expected, would glow under fluorescent light, enhancing his use for research. In fact, ANDi failed to glow. But there followed a series of experiments creating mice, fish, plants, and bacteria as artificial organisms that do glow in the dark (Fig. 10). They are regarded less as creatures than as "living image-generating apparatuses."[32]

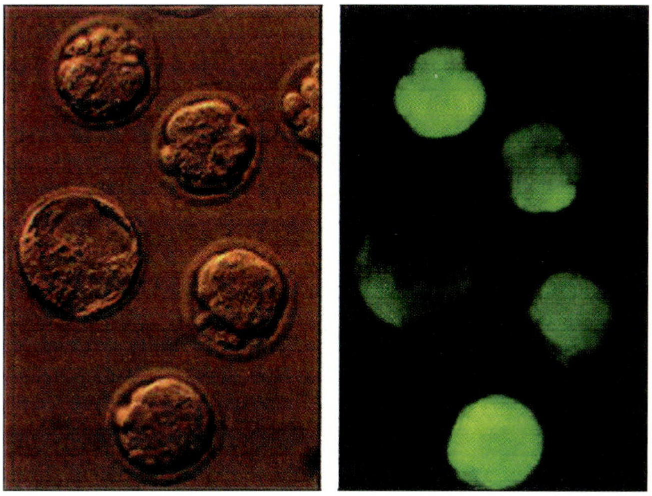

8 Phosphorescent diagnostic cells.

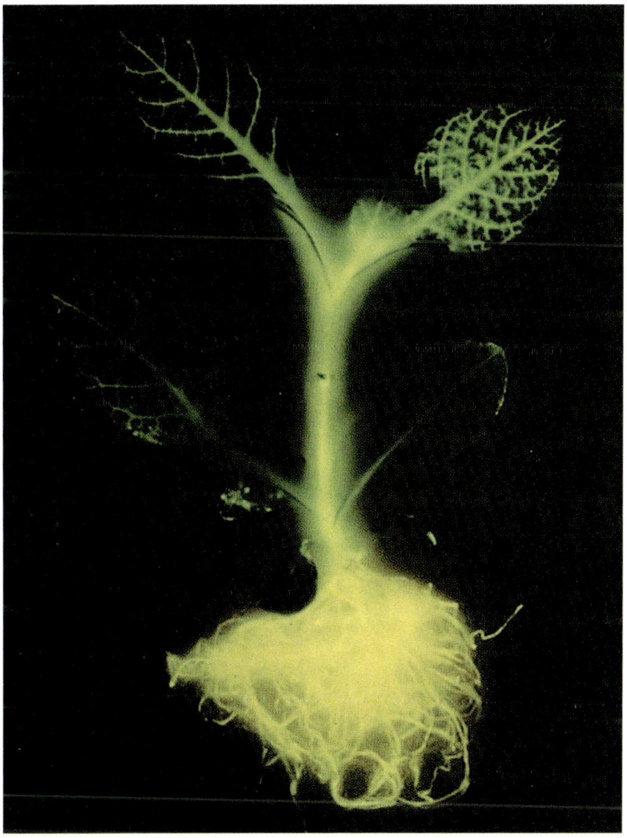

9 David Ow, Tobacco seed combined with firefly genes.

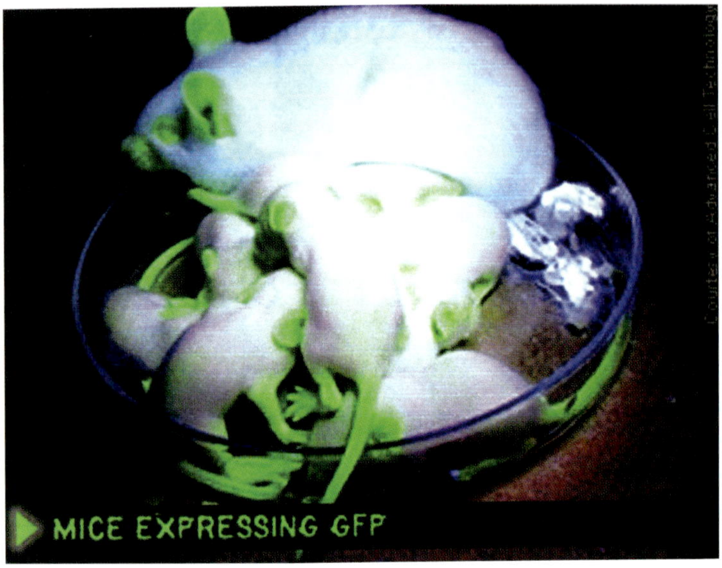

10 Glow-in-the-dark mice.

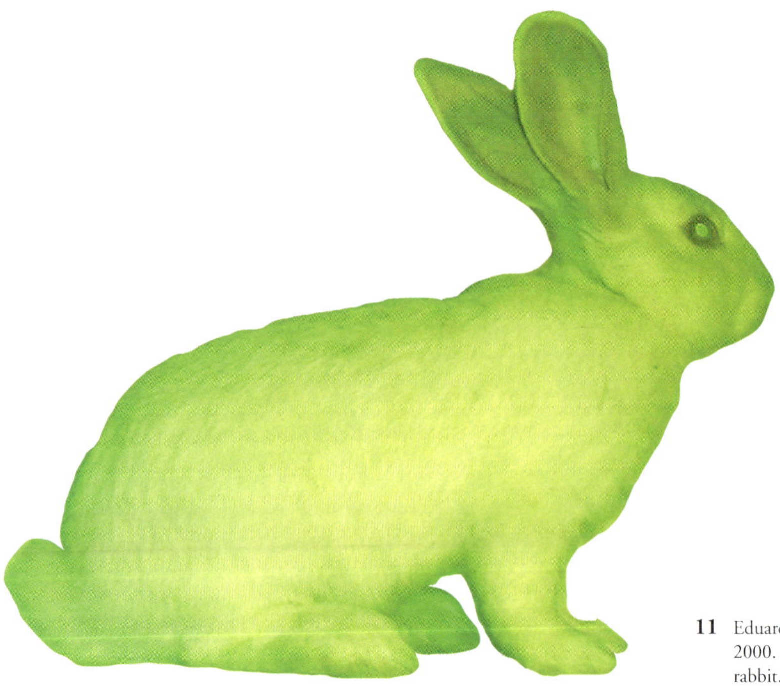

11 Eduardo Kac, *GFP Bunny,* 2000. Alba, the fluorescent rabbit.

Artist Eduardo Kac's *GFP Bunny* (2000), Alba (Fig. 11), is an art-world evocation of the bioluminescence experiments that created ANDi, the transgenic monkey, and more recent efforts to inject luminescence into mice. Kac worked with French geneticist Louis-Marie Houdebine to create his transgenic bunny, which is allegedly luminous and green from the injection of fluorescent GFP from jellyfish.[33] When Kac presents photos of his transgenic work to the public, Alba appears as bright chartreuse. However, for Alba to actually glow, she would have to be shaved and then exposed to blue light with a wavelength of 488 nanometers. To actually create such a bioluminescent creature would require technologies that exist only under highly special laboratory conditions. Some suggest that Alba was a Photoshop construction, but Kac maintains that the photograph was not altered. Meanwhile, Alba met an untimely death in equally vague circumstances.

Transgenic Art

Eduardo Kac explicitly defines himself as a transgenic artist. Transgenic art, he writes, "is a new art form based on the use of genetic engineering techniques to transfer synthetic genes to an organism or to transfer natural genetic material from one species into another to create unique living things."[34] He proposes that artists appropriate techniques of genetic engineering to invent new life forms. This would expand the conceptual boundaries of art and, he suggests, it may also stimulate a useful dialog about the ethical and social implications of genetic research. The intention of artists who have altered biology in their work is not the same as those of scientists or biotechnology labs. But his kind of art itself creates ethical dilemmas as artists manipulate living things without the oversight and constraints routinely imposed upon scientific experiments. Cultural historian Edward Shanken, however, states that genetic engineering performed by artists "interrogates the limits of knowledge and consciousness to plumb the depths of the human condition." He sees this as a socially significant aspect of transgenic art.[35]

The art collective from Australia known as Tissue Culture and Art Project is composed of Oron Catts, Ionat Zurr, and Guy Ben-Ary. The threesome employ living tissue as their sculptural medium. In a recent exhibition, *Pig Wings Project* (2000–2001) (Fig. 12) at the de Cordova Museum in Lincoln, Massachusetts, the artists collaborated with Dr. Joseph P. Vacanti to create three types of wings: bird, bat, and reptile, from pig cells.[36] Dr. Vacanti, director of the Tissue Engineering and Organ Fabrication Laboratory at Massachusetts General Hospital, and his laboratory partners had formerly worked on the growth of a human-size ear on the back of a mouse (Fig. 13). Dr. Vacanti considers the work of these artists to be "living art," that "will challenge our concepts of the living world versus the nonliving."[37]

Joe Davis' *MicroVenus*, exhibited at Ars Electronica in 2000, is a molecular artwork comprising a synthetic DNA molecule. Housed in a biological

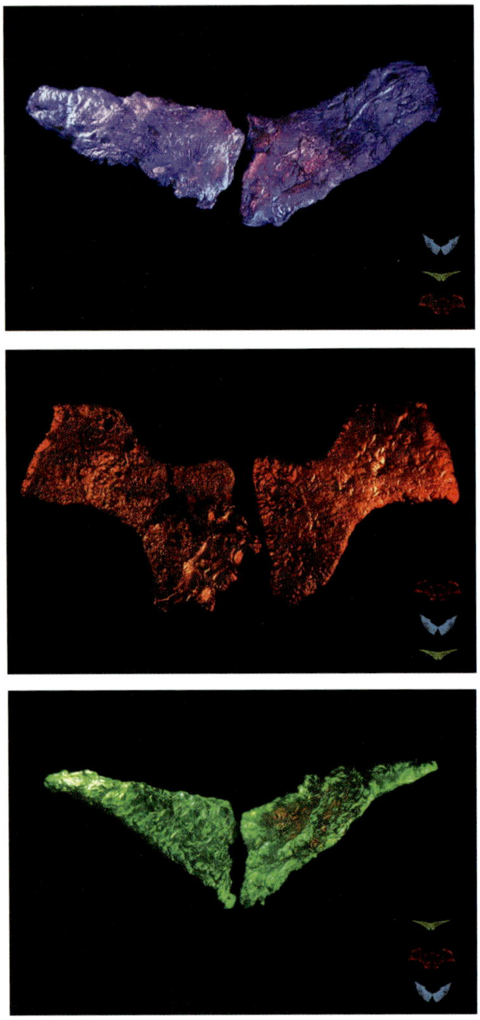

12 Tissue Culture and Art Project. *Pig Wings Project: The Chiropteran Version, the Aves Version, and the Pterosaurs Version,* 2000–2001.

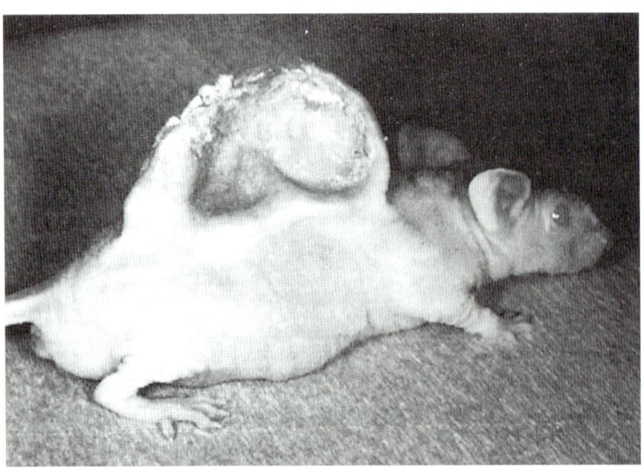

13 Mouse with tissue-engineered cartilage in shape of human ear.

containment apparatus, this genetically engineered bacterium was shown for the first time in Linz, Austria.[38] David Kremers, Caltech's Distinguished Conceptual Artist in Biology, employs genetically engineered bacteria to produce colored enzymes that simulate a self-directed painting machine. Laura Cinti has created a cactus that grows human hair.[39]

Jump Cut: Contemporary Art and the Chimeric Image

In the tradition of Dada and surrealism, contemporary artists employ the cut-and-paste techniques of collage to comment on the creation of transgenic creatures. Grünfeld's *Misfit* is but one example. Installation and performance artist Paul McCarthy refers to transgenic art as part of his interest in developing "a repository of the fears, obsessions, and conflicts that face the human species at an evolutionary crossroads."[40] McCarthy's work includes hybrid mutants such as his 1994 installation, *Tomato Heads* (1994) (Fig. 14), which can be considered as a takeoff on the bioengineered tomato that the media referred to as "Frankenfruit." Through this work, McCarthy questions the validity of assumed values, fixed identities, and clear-cut boundaries. "I am interested," he writes, "in things having multiple identities." He explicitly draws examples from DNA research.[41]

Wood sculptor Jane Schneider captures the concept of changing categories in a 2003 exhibit titled *Mutables*, a body of work that alludes to lifelike objects capable of undergoing transformation.[42] Her abstract sculptural installation, *What's That?* (2002) (Fig. 15), is made from cedar root systems suspended by steel cables from the ceiling. Imagistically part human and part animal, her sculptures elucidate unsettling discontinuities involved in the creation of transgenic artifacts. In such transgressions, physical distinctions between animal and human forms are blurred. "There is no longer a given mold for ordinary objects," says the sculptor.

In his lavishly produced five-part epic film, *Cremaster Cycle*, Matthew Barney exploits the myth of the satyr. In Greek mythology, the satyr is part man and part goat and is possessed with a lecherous sexual appetite. In *CREMASTER 4* (1994) (Fig. 16), Barney is seen decked out with elongated animal ears as he masquerades as a man–animal conjunction. Underscoring Barney's interest in biological models is the ambiguity inherent in embryonic developments of gender and identity as a transmutable act of becoming.[43]

Gerhard Lang's *Palaeanthropical Physiognomy* (1991/2000) (Fig. 17) is an imaginary zoological menagerie related to 19th-century classification systems in natural history, anthropology, and criminology. Transforming human faces into beastly composites, Lang's portraits, too, are reminiscent of le Brun's (see Fig. 3, page 12). However, this contemporary German artist's work is purely technologically based. In addition to photography and light projections, he employs the Indentikit. The Indentikit, a composite picture-

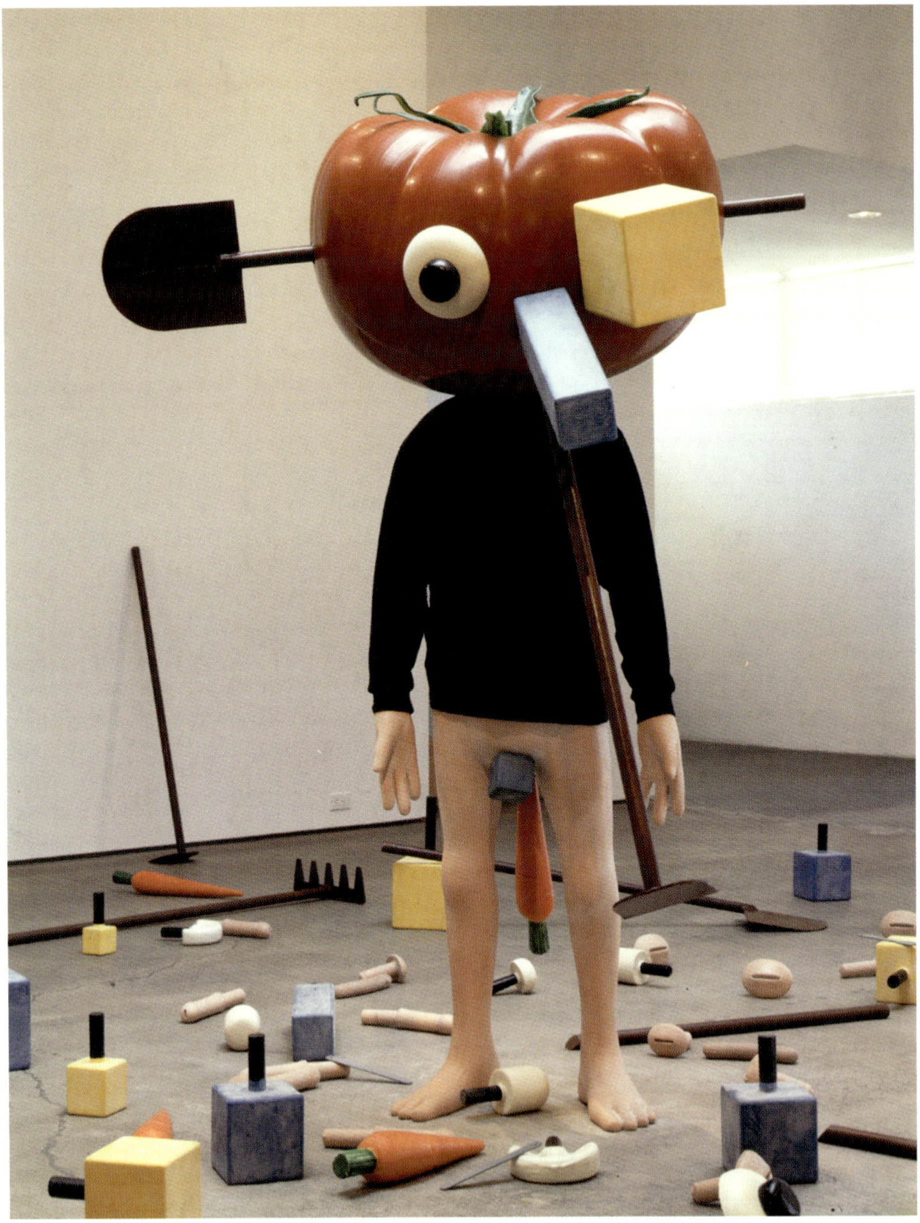

14 Paul McCarthy, *Tomato Heads* (detail), 1994. Courtesy of the artist and Luhring Augustine, New York.

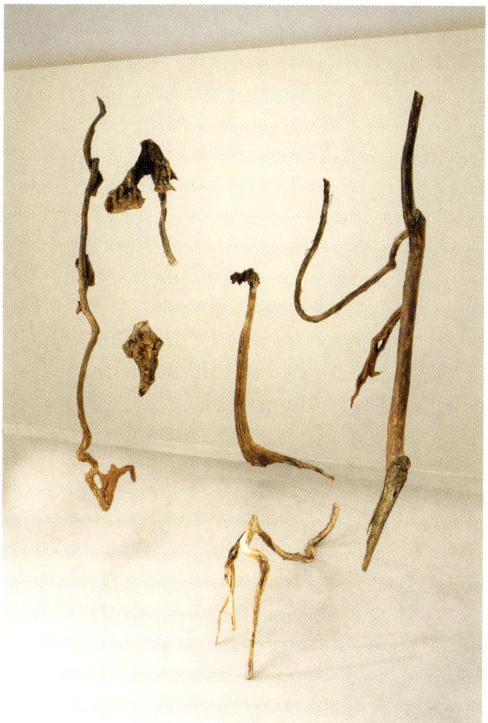

15 Jane Schneider, *What's That?* 2002.

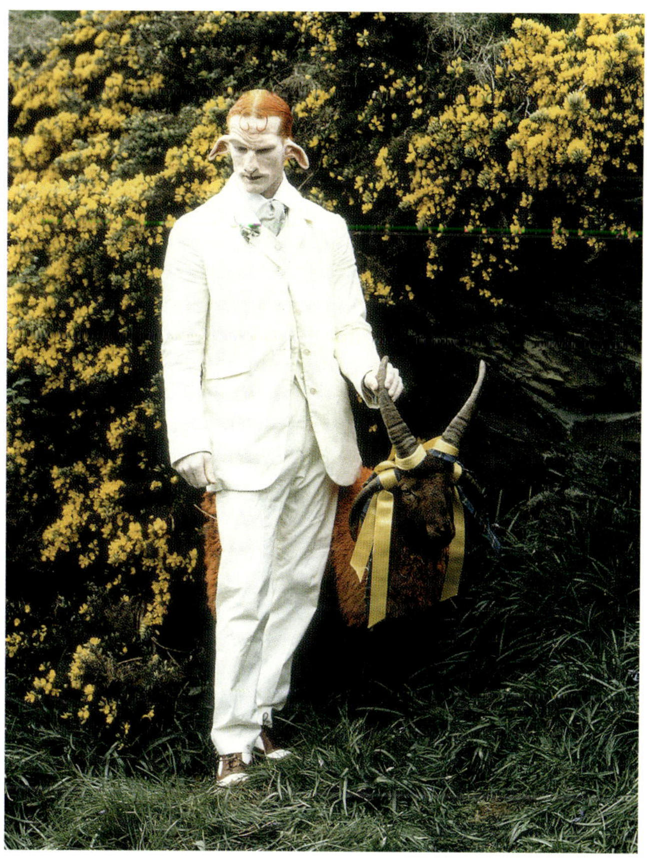

16 Matthew Barney, *CREMASTER 4*, 1994.

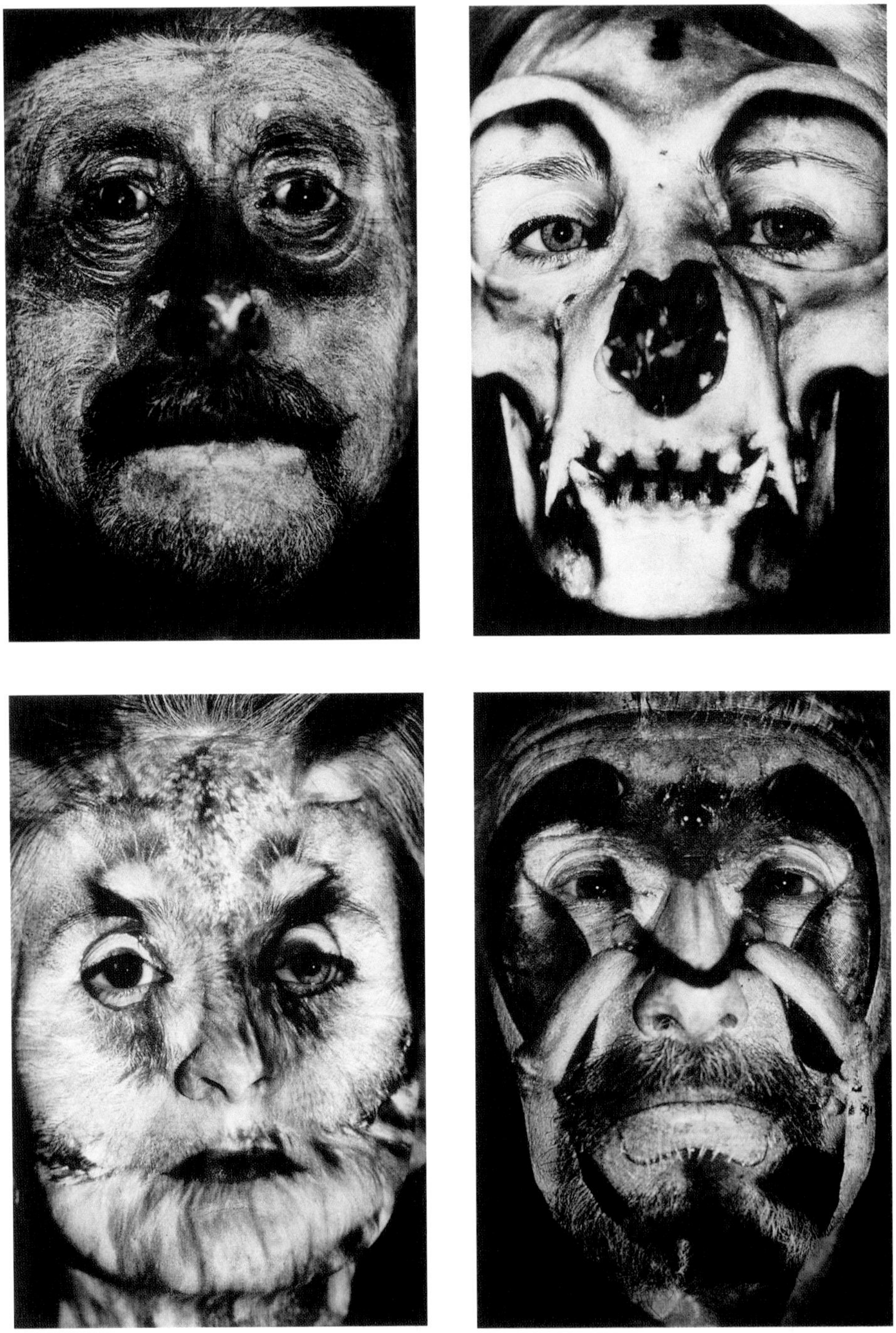

17 Gerhard Lang, *Palaeanthropical Physiognomy* (detail), 1991/2000.

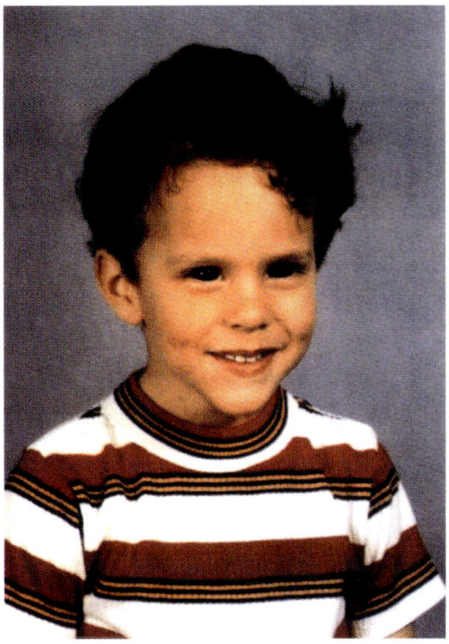

18 Bradley Rubenstein, *Untitled (Boy with Puppy-dog eyes)*, 1994–1995.

making apparatus, was initially produced by Minolta in the 1950s and 1960s as a visualizing tool to reconstruct the probable faces of Hiroshima victims.[44] This method was also used by the German police to help identify criminals. Working with a criminologist, Lang utilized what he called "his passport collection of photos of beetles, bees, owls and primates" to create his montages.[45] These images were then projected over skulls and faces for the further blurring of boundaries. Tribal looking, but instrumentally based, these images confront the viewer with the idea of evolutionary time from the primitive to the modern in both art and science.

In a series that initially seems less disturbing, Bradley Rubenstein delicately postures young children who are, in fact, joint animal–human creatures (Fig. 18). At first glance, in *Untitled (Boy with Puppy-dog Eyes)* (1994–1995), a young schoolboy sits and smiles as he ostensibly poses for his class picture. Upon close investigation, however, this boy is no ordinary child. Through the manipulations of the artist—employing Photoshop techniques—the boy's eyes take on the deep and warm characteristic of a puppy dog's. Rubenstein's work, conceptually similar to Lang's, suggests how the boundaries between human and animal are becoming more fluid and often hard to explicate, if not discern.

Images of transgenic animals in Alexis Rockman's *Biosphere: Laboratory* (1993) (Fig. 19) recall H.G. Wells' beast people on the *Island of Dr. Moreau.* This large-scale painting portrays a futuristic setting in an orbiting outer-space sphere. Genetically altered mutant beasts are strewn about a laboratory floor. Animals in cages await their fate. Like Wells, Rockman uses chimeras to question the consequences of scientific experiments. In another painting, *The Farm* (2000) (Fig. 20), Rockman portrays a number of trans-

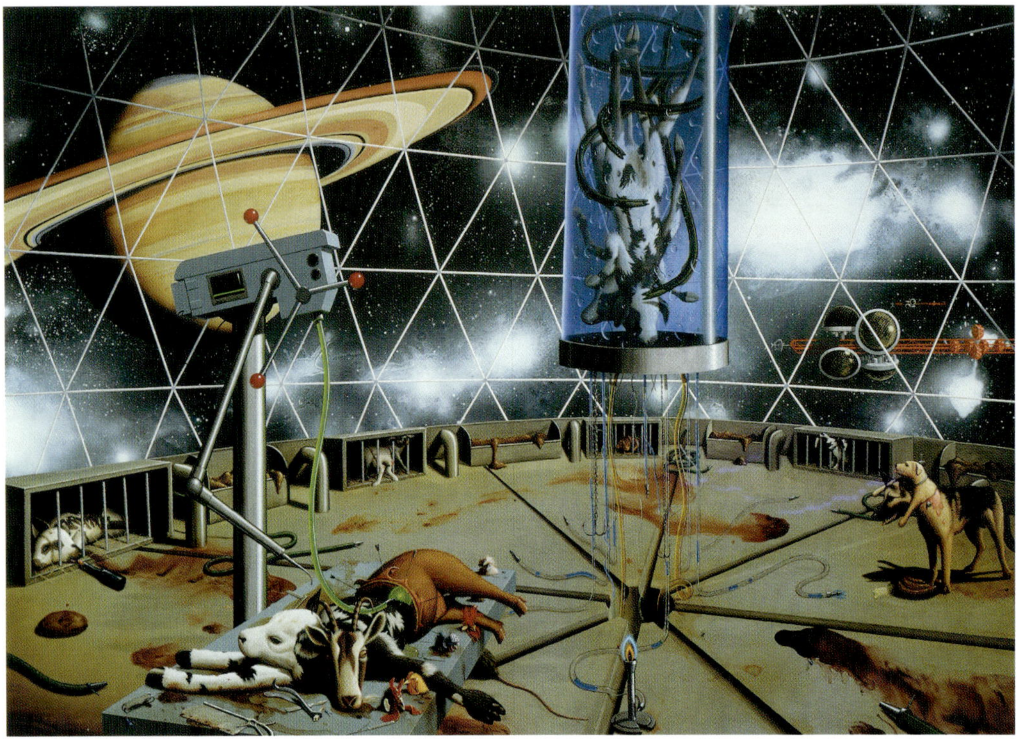

19 Alexis Rockman, *Biosphere: Laboratory*, 1993.

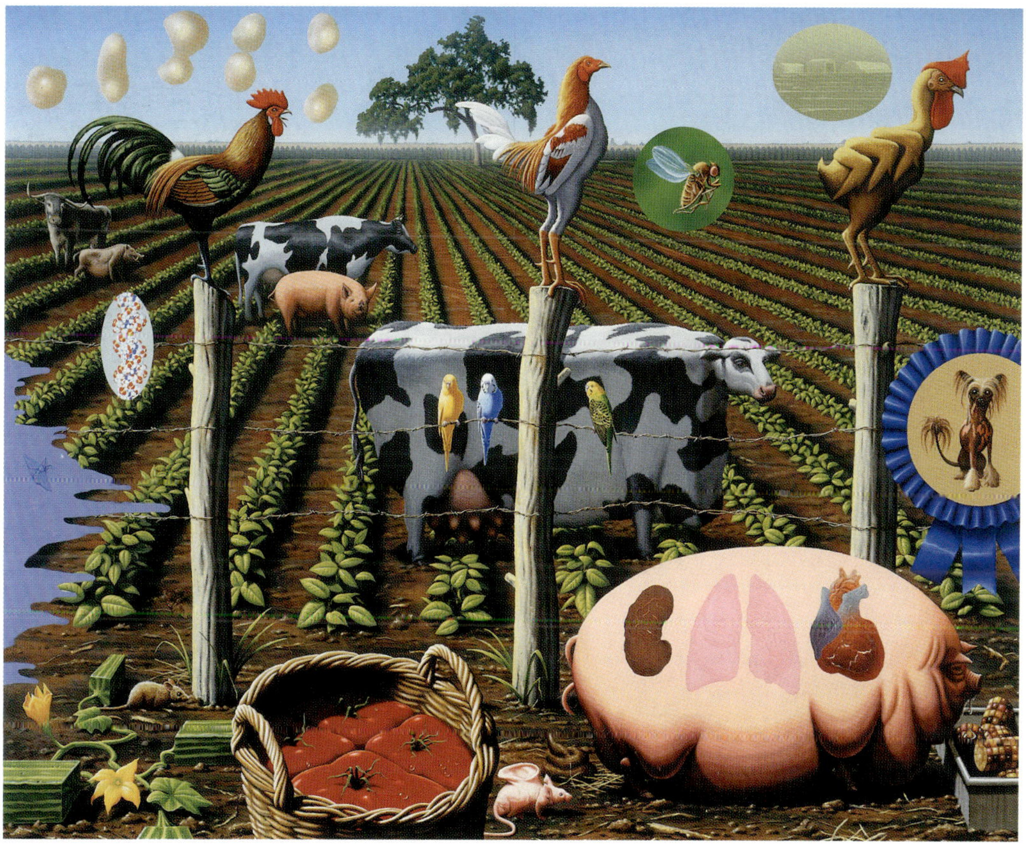

20 Alexis Rockman, *The Farm,* 2000.

genic species—a cow with a geometric body shape, a pig being farmed for its organ parts, and a mouse with a human ear. This painting plays off the visually bizarre experiments in which genetically engineered mice have human ears (see Fig. 13).

Transgenics and Cyborgs

The creation of transgenic creatures is but one example of the blurring and blending of boundaries in science and technology today. Traditional boundaries have been besieged on many fronts.[46] Cultural critic Andrew Ross describes a whole range of "genetic overhauls and cybernetic enhancements—boosterware, biochip wetware, cyberoptics, bioplastic circuitry, designer drugs, nerve amplifiers, prosthetic limbs and organs, memoryware, neural interface plugs and the like." He calls the body "a switching system, with no purely organic identity...."[47]

The "cyborg," a term coined in the 1960s to describe a cybernetic organism, has, in Donna Haraway's words, come to symbolize "fractured identities, the blurred and anxiety-inducing boundaries between human and animal, and between organism and machine."[48] Artists, long attracted to the border between organic and constructed works, problematize these boundaries with the range of their own techniques.

For example, Alan Rath's robotic sculptures appear to possess fleshy human attributes. They display digitalized moving images of isolated body parts, rolling eyes, lapping tongues, and audio speakers whose pulsations suggest a beating heart.[49] Scientific robot projects are frequently included in art exhibitions. Robo-sapiens had become a generic term for robots that assume a human form and human attributes and even seem to express emotion....[50] Just as the dividing lines between persons and robots are blurred in this work, so too are the boundaries between science and art.

Frank Gillette, using computers, generates visions that meld the organic with the biological, resulting in what he calls "fictional naturalism," as in *HCE Series* (1999–2000) (Fig. 21) and *A Tangled Bank* (2000) (Fig. 22). Beginning with images from the real world, this media artist pushes and pulls, morphs and layers, distorts and confounds the ready-made. Through programmatic manipulation of digital tools Gillette's visual domain is inhabited by, among other things, the chimera. In subtle transformations, animal parts only give up their recognizability after the viewer has invested some time observing. Playing hide and seek, and harkening back to Max Ernst, Gillette's creatures emerge from a kind of digital and stochastic Darwinism, evolving by chance, design, and whimsical choice.

Eva Sutton's interactive digital installation, *Hybrids* (2000) (Fig. 23), invites viewers themselves to alter the combination of animal parts through

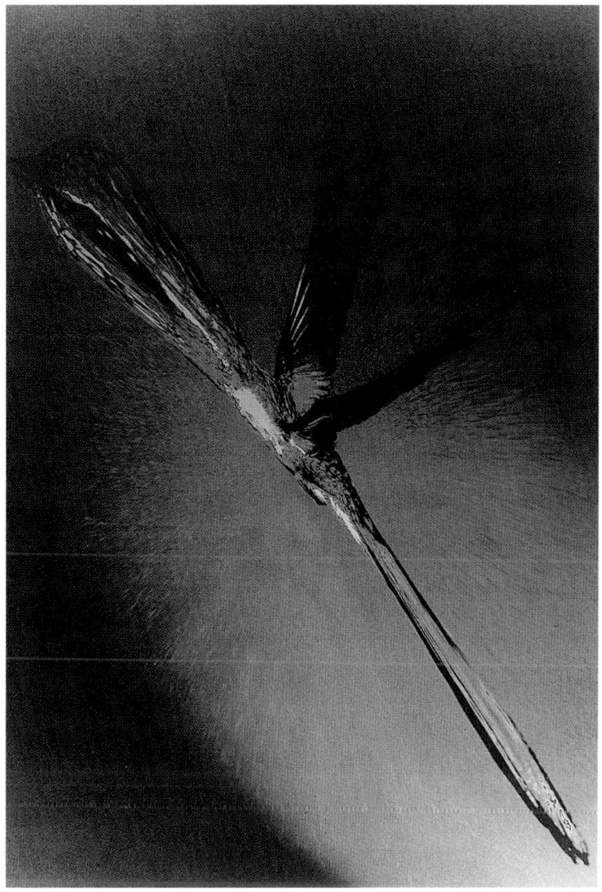

21 Frank Gillette, from the *HCE Series,* 1999–2000.

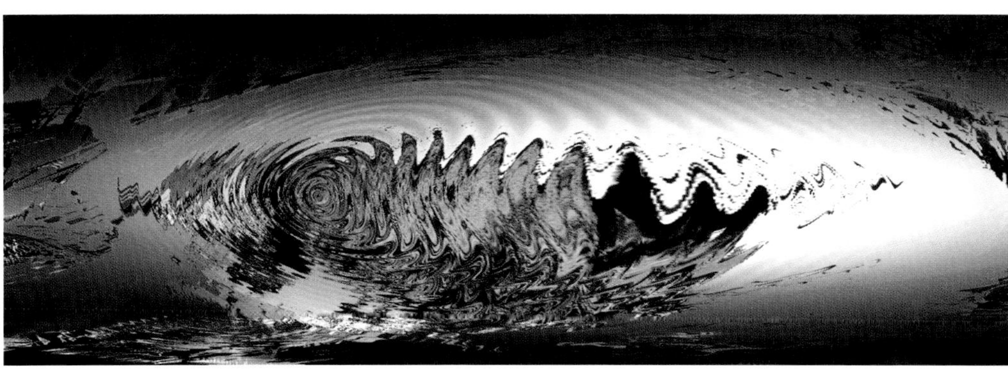

22 Frank Gillette, *A Tangled Bank,* 2000.

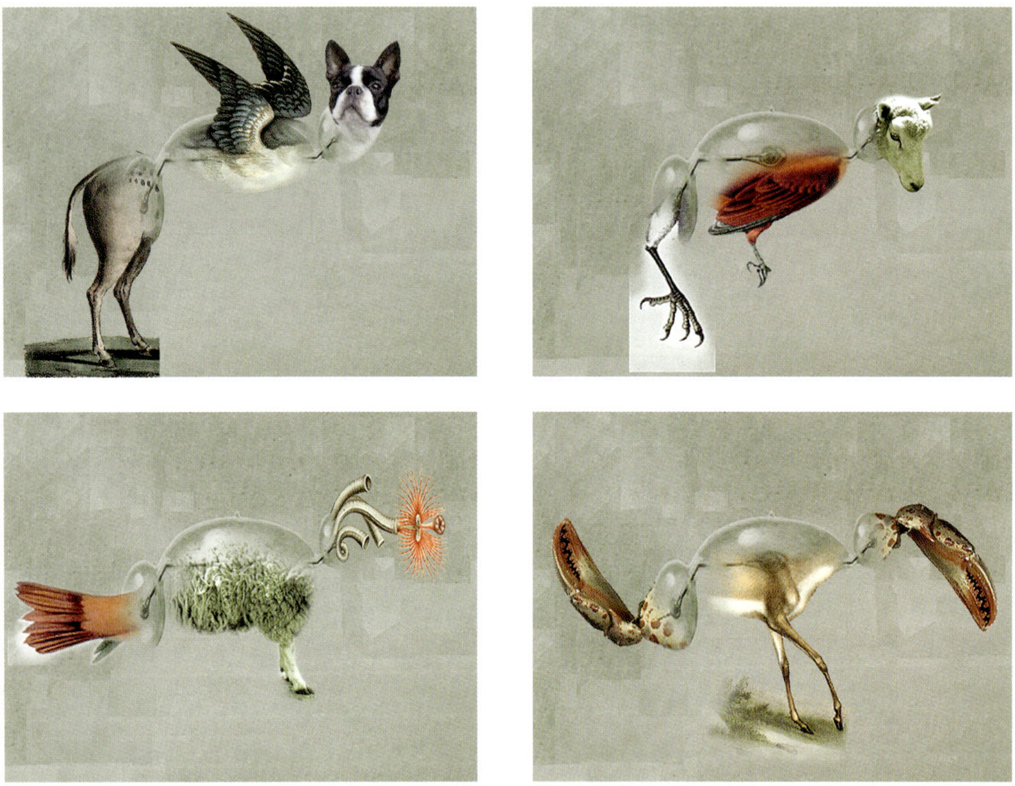

23 Eva Sutton, *Hybrids* (detail)
2000.

a process of random selection, a kind of game that is played with unpredictable results. Sutton had worked as a software designer of scientific instruments. She became aware that modular thinking involved in software design was also being applied to biological systems and that modular components of the body were actually being manipulated much like software systems. "Systems that we can imagine—be they perfect strawberries, perfect children, perfect babies—can be designed and built. But what are the consequences of achieving perfection?"[51] Her response, in *Hybrids,* driven by the monsters in Grimm's stories that were read to her as a child, turns fairy tale into fact.[52] She describes *Hybrids* as "Animal heads, torsos, and hind legs juxtaposed at a wildly varied scale [in which] the images flip like a surrealist slot machine."

Aziz + Cucher are a collaborative team whose work has been described as a "series of organs without a body.[53] Their deviant figures (Fig. 24) are devoid of any recognizable body parts. Rather, they appear as quasi-geometric creatures enveloped in skin and stripped of all detail. The work convincingly conveys the ambivalent poetics of the biotechnological world. On the one hand, Aziz + Cucher are creating unidentifiable mutants, but they are as well celebrating the intoxicating thrill of technological hyperprogress. Along parallel lines, Marcia Lyons' *Munging Bodies* (1996–present) are mutated computer images of eroticized human flesh. These body parts will perhaps one day come in all shapes and sizes. And for Patricia Piccinini, Australia's representative to this year's Venice Biennale, the interplay with the "natural" and synthetic has been a theme in forming her work. Most recently she has created SO_2, an imaginary animal creation taking a cue from SO_1, the first synthetic laboratory organism.

Such digital techniques are also ripe for social satire. In an on-line game, artists Natalie Bookchin and Jin Lee have created a digital chimera called "MetaPet" in which cyber-humans are implanted with genes for canine obedience. Players assume the role of corporate managers in biotechnology firms and they are given the task of making transgenic humans as productive and efficient as possible at the very least cost.[54]

Other futuristic scenarios are envisioned by techno-artists, architects, and software creators.[55] William Latham, Karl Sims, Greg Lyn, and Rebecca Allen use computer programs to develop visual systems that mutate or animate live forms. From issues concerning emergence, evolution, artificial intelligence, and bio-architecture, these practitioners interface with industry, the art and design world, and the motion picture industry.[56]

Boundary Transgressions in Social Context

Chimeras once populated the literature, mythology, and art of ancient Greece. They were dangerous, formidable, and powerful beasts, representing

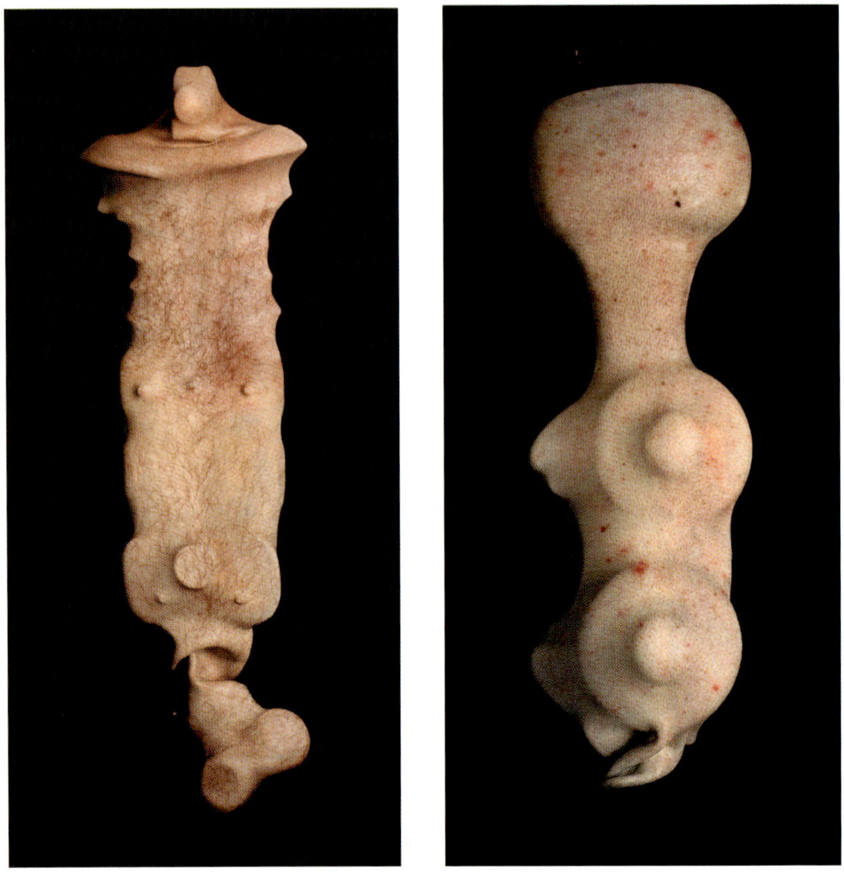

24 Aziz + Cucher *(left) Chimera #1,* 1998; *(right) Chimera #2,* 1998.

fantastic yet uncivilized and chaotic forces in nature that confronted mankind. But why is there a revival of transgenic images in visual art today? In some cases, they are a direct response to technologies that appear to threaten the very integrity of species. In other cases, they represent the ability to extend and transform life. These visual images of hybrid forms may also represent a model of the modern mind—a fabrication of the chaos, discordance, and incongruity that seem to characterize contemporary culture. Perhaps the fragmented and reconstituted body has become, as Linda Nochlin suggests, a fundamental metaphor of postmodernity.[57]

The chimera must be understood in a social context: These beasts have always reflected the social, scientific, and religious circumstances of their time.[58] Many of the transgenic experiments of science today are, to say the very least, controversial—the subject of political, religious, and ethical debate. Just as H.G. Wells' "beast people" were a cynical commentary on the scientific manipulations that were taking place in the 19th century, so Rockman's laboratory animals portray a horror of experiments in genetic engineering taking place in today's laboratory. Other artists, particularly the media artists, architects, and designers, look toward these biological models as experimental sources bringing forth new biomorphic forms and materials. Some artists, like Eduardo Kac and Paul McCarthy, make explicit reference to transgenic experiments as having influenced their work. Others, like Frank Gillette and Eva Sutton, create transgenic images, as a way to represent the wider transgression of boundaries in the postmodern world.

Boundaries and borders are conjunctions of danger; boundary transgressions violate traditional definitions and defy assumptions about the apparently rational. But in effect, novel technologies create "new" rational orders. Some contemporary art reflects a certain amusement at defying rationality in this way, but artists are also expressing popular anxieties about the threats to authenticity and species integrity that result from some of the most striking research in contemporary science.

Notes

1. This was Homer's description in the 9th century BC and the common dictionary description today, but there are other variations of the Chimera. The Kimaira who often appears on ancient Greek pottery is visualized in Hesiod's version with three heads.
2. Chimeras can be regarded as a subset of freaks. Under the word monster, in its third definition, the OED defines such entities as "imaginary animal, part human" or "parts of different animals."
3. Leslie Fiedler. *Freaks: Myths and images of the secret self.* Anchor Books, New York, 1993, p. 168.
4. Jan Bondeson. *A Cabinet of medical curiosities.* W.W. Norton, New York, 1997, pp.144–157.
5. Bondeson, op. cit., p. 170.
6. Bill Bynum. 2002. Maternal impressions. *Lancet* 359: 898. Interpretations of Merrick's disease have changed over time. Originally, he was thought to suffer from elephantiasis (a transmitted disease), then neurofibromatosis (a genetic disease), and more recently, Proteus Syndrome (a non-inherited condition).
7. William T. Alderson, ed. *Mermaids, mummies and mastodons: The emergence of the American museum.* American Association of Museums, Washington, D.C., 1992, p. 34.
8. H.G. Wells. *The island of Dr. Moreau.* (1896). Reprinted by Dover Publications, New York, 1996, p. 62.
9. Jon Turney. *Frankenstein's footsteps: Science, genetics and popular culture.* Yale University Press, New Haven, 1998, pp. 55–56.
10. Mikhail Bulgakov. *Heart of a dog.* Grove Press, New York, 1968.
11. Suzanne Anker. Material into metaphor: Picturing DNA. Paper delivered at the London Network for Modern Fiction Studies, May 27, 2000.
12. Dada, an international art movement in the early 20th century, and its later codification as surrealism generated provocative art that mocked traditional representa-

tions and values. See Richard Huelsenbeck, ed. *DADA Almanac.* Atlas Press, London, 1993.

13. Mark Dery. *The Pyrotechnic Insanitarium.* Grove Press, New York, 1999.
14. William Rubin. *Dada, surrealism, and their heritage.* The Museum of Modern Art, New York, 1968, p. 50.
15. Werner Spies. *Max Ernst collages: The invention of the surrealist universe.* Abrams, New York, 1968.
16. Bruce Altshuler. Dada is Politsch. In *The avant-garde in exhibition: New art in the 20th century.* Abrams, New York, 1994.
17. Barry Commoner uses the metaphor of montage to describe a concept of "alternative splicing" that, he claims, undermines the central theory of Watson and Crick: "In alternative splicing, the gene's original nucleotide sequence is split into fragments that are recombined in different ways to encode a multiplicity of proteins, each of them differing in their amino acid sequence from each other and from the sequence that the original gene, if left intact, would encode." Barry Commoner. 2002. Unraveling the DNA myth. *Harper's Magazine* February, p. 42.
18. Technically, chimeras and transgenics are different categories of human/animal inventions. A chimera is composed of cells from two genetically distinct embryos. Chimeras can be created by introducing embryonic cells (or stem cells) into a host embryo. They then become "progenitor cells" that mature to replace equivalent host cells. However, each individual cell contains the genetic material from only one of the contributing organisms, so the characteristics of the chimera are not passed on to the progeny. In contrast, transgenic animals and plants are created by introducing a foreign gene into a fertilized zygote. Because the foreign gene is incorporated into the host's genome at the one-cell stage, the transgenic organism has foreign DNA in every cell. Thus, the characteristics of a transgenic organism can be passed on to the next generation.
19. For an anthropological analysis of transgenics, see Donna Haraway. *Feminism and technoscience. Modest-witness@second-millennium FemaleMan-meets oncomouse.* Routledge, New York, 1997.
20. Nicholas Wade. 2002. Stem cell mixing may form a human-mouse hybrid. *The New York Times,* November 27, p. A21.
21. Associated Press. 1998. Mouse grows elephant egg in test. *The New York Times* November 10. This was predicted by journalist Arthur Lubow in 1977: "The DNA of mice and elephants can now be joined in bacteria." See Turney, op. cit., p. 195.
22. Jennifer Couzin. 2002. Wanted: Pig transplants that work. *Science* 295: 1008.
23. David Cooper and Robert Lanza. *Xeno: The promise of transplanting animal organs into humans.* Oxford University Press, New York, 2000.
24. Robert Budd. 1999. Biotechnology in the twentieth century. *Social Studies Sci.* 21: 415–457.
25. US Congress, House of Representatives Committee on the Judiciary, Subcommittee on Courts, Hearings on "Patents and the Constitution: Transgenic Animals," 100th Congress, 1st session, June 1998.
26. Turney, op. cit., p. 207.
27. Interestingly, in the original comic book of the 1970s, Spiderman had been bitten by a spider exposed to radiation, suggesting how popular culture reflects the anxieties of a given time.
28. Gretchen Vogel. 2001. Transgenic animals. Infant monkey carries jellyfish gene. *Science:* 291: 226.
29. Steven Levingston, *International Herald Tribune,* January 18, 2001.

30. Mike Toner, *The Atlanta Journal and Constitution*, April 2, 1994.
31. Ow D.W., Wood K.V., DeLuca M., de Wet J.R., Helinski D.R., and Howell S.H. 1986. Transient and stable expression of the firefly luciferase in plant cells and transgenic plants. *Science* 234: 856–859. At http://www.pgec.usda.gov/Ow/Owpubs.html
32. According to Reinhard Nestelbacker, an Austrian molecular biologist. Ars Electronica, Linz, Sept 1-6, 2001.
33. David Joselit. 2000. Biocollage. Carol Becker, "GFP bunny." *Art J.* 59: 44.
34. Leonardo Electronic Almanac 6, 11, 1998. At: http://mitpress2.mit.edu/e-journals/LEA/LEA2002/LEA/index.htm
35. Symposium on "Art, Ethics and Genetic Engineering: The Transgenic Art of Eduardo Kac," Duke University, November 6, 2000.
36. Ann Wilson Lloyd. 2003. Pigs won't fly, but questions may. *The New York Times*, March 16th.
37. Ibid. Also see SymbioticA, "a research laboratory dedicated to the artistic exploration of scientific knowledge in general and biological technologies in particular. It is located in The School of Anatomy and Human Biology at the University of Western Australia." http://www.symbiotica.uwa.edu.au.
38. W. Wayt Gibbs. 2001. Art as a form of life. Profile. *Sci. Am.* 284: 40–41. At: http://www.sciam.com
39. http://www.lauracinti.com. Other artists working with "live DNA" include Brandon Ballengee, Adam Zeretsky, Heather Ackroyd, Dan Harvey, and Natalie Jeremijenko. Also see Eduardo Kac (ed.), *Biotechnology, art and culture.* MIT Press (in press).
40. Dan Cameron. Paul McCarthy. Brochure for exhibition at New Museum of Contemporary Art, February 24–May 17, 2001.
41. Fereshteh Daftari, interview with Paul McCarthy in the brochure for his exhibition at The Museum of Modern Art, June 1–July 18.
42. Exhibition at the Maris Gallery in Poughkeepsie, New York, 2003.
43. For a full discussion of Barney's work, see Nancy Spector. *Matthew Barney: The CREMASTER Cycle.* Guggenheim Museum, New York, 2002.
44. Lucius Burkhardt. "Palaenthropical Physiognomy" (catalogue), Frankfurt, 1993.
45. Interview with the artist in MINED, TANK Publications, London, 2002.
46. Examples are numerous in medical contexts, and in the field of artificial intelligence where human experience is reduced to the stimulation of the neo-cortex. Heart valves, pacemakers, and prosthetics are routinely transplanted into human bodies. Implants simulate hearing for the deaf, implanted neuro-stimulators send electric charges to alleviate pain. Researchers are developing a computer, called "soul catcher," that can be implanted in the brain to complement human memory and gather data transmitted by wireless networking. As artificial organs become the frontier of technocratic medicine, the body is re-imagined as a machine, a hybrid form, indeed a kind of chimera that morphs the human and the machine.
47. Andrew Ross. *Strange weather.* Versus, London, 1991, pp. 152–153.
48. Donna Haraway. *Simians, cyborgs, and women: The re-invention of nature.* Routledge, New York. 1991.
49. Worcester Art Museum, Alan Rath, press release.
50. Peter Menzel and Faith D'Aluisio. *RoboSapiens: Evolution of a new species.* MIT Press, Cambridge, 2000.
51. Eva Sutton, interview in catalog of Paradise Now.
52. Eva Sutton. 2001. Thoughts on hybrids: Intention and implementation. *ArtLab* 23: Issue 1, Spring. At: http://www.artlab23.net

53. *Aziz and Kucher: Recent works 1998-9.* Exhibition Catalog from Galerie Lutz Teutloff, Beilefeld, Germany, Catalog essay by Thyrza Nichols Goodeve.
54. http://metapet.net
55. Steven Henry Madoff. Out of the ether, a new continent of art. *The New York Times,* February 14, 1999.
56. Steven Henry Madoff. The wonders of genetics breed a new art. *The New York Times,* May 26, 2002. See http://www.nemeton/axis-mutatis/latham.html, http://www.Genarts.com/kar/genetic-images.html, http://www.glform.com, http://emergence.design.ucla.edu/people/biorebal.htm.
57. See Linda Nochlin. *The body in pieces: The body as a metaphor of modernity.* Thames and Hudson, London, 1994.
58. Lorraine Daston and Katherine Park. *Wonders of the order of nature.* Zone Books, New York, 2001, p. 191.

Breeding Better Babies: A New Eugenics?

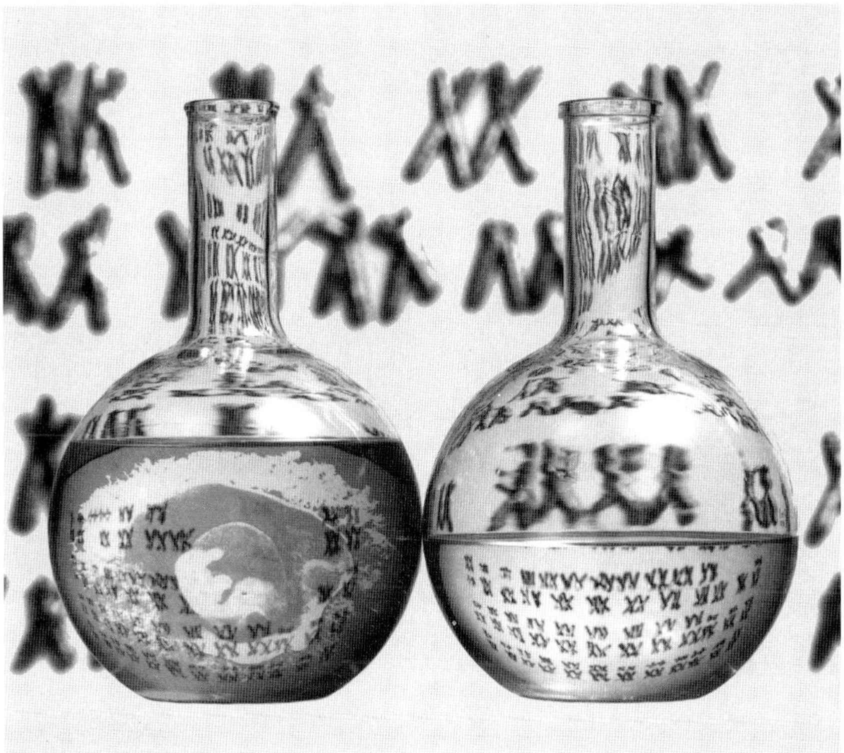

1 Suzanne Anker, *Material Powers,* 1999.

SUZANNE ANKER'S *MATERIAL POWERS* (1999) (Fig. 1) portrays a fetus afloat in a flask. Surrounded by hovering shadows of chromosomes, this "baby in a bottle" is a scheme for an emboldened future. A second flask awaits its fetal implant. Through techniques of assisted reproduction involving in vitro fertilization, frozen embryo transfer, and genetic engineering, babies can be bred by design. Anker's visualization of a glass-encased fetus addresses the issues arising from assisted reproductive technologies, commonly (and ironically) referred to by the acronym ART. Her suspended tiny creature, with its chromosomal icons, prys open questions addressing the potential of eugenic practices in an age of reproductive technology. Will artificial wombs join petri dishes and test tubes as yet another environment for developing, propagating, and manipulating life?

The vision of babies in bottles has long captured the cultural imagination. Historian Susan Squier traces images of babies in bottles from the early days of embryonic biology. These include Charles Kingsley's 1863 children's story, *The Water Babies* (which had an illustration of Thomas Huxley, at right in figure, examining a bottled water baby) (Fig. 2), and Julian Huxley's ruminations during the 1920s about human development and differentiation.[1] The first science fiction magazine, *Amazing Stories,* featured futuristic stories about alternative methods of reproduction throughout the 1920s. In 1927, the magazine included a reprint of Julian Huxley's "Tissue Culture King" as well as A. Hyatt Verrill's "The Ultra Elixir of Youth" and Francis Flagg's "The Machine Man of Ardatha." In *Brave New World* (1932), Aldous Huxley described a society in which people are born—that is, hatched—in ectogenic vessels and then mass produced. These fictional works reflected disquieting concerns about ectogenesis and those problems posed by scientific intervention into the origin of life.[2]

Creating Life in the Lab

Experimental studies of cell activity in the 1930s, such as those conducted at the Strangeways Laboratory in Cambridge, UK, were believed to be the first steps toward creating life in the lab. Journalists worried about "a brave new world of babies cultured in test tubes."[3] "Will the sexless soulless crea-

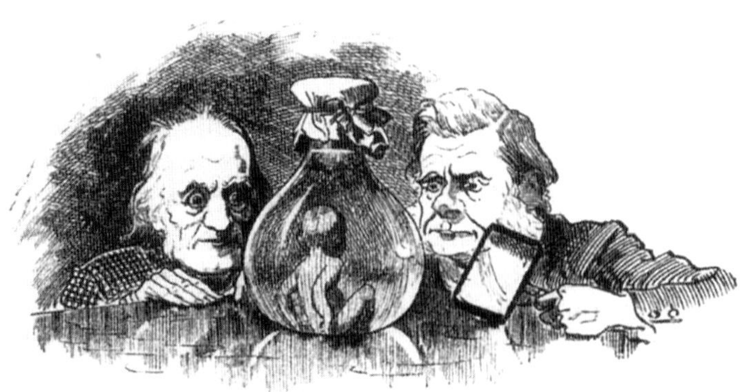

2 Charles Kingsley's illustration for *The Water Babies,* 1863.

tures of chemistry conquer the true human beings?" The archives of the Strangeways Laboratory contain an anonymous poem about this brave new world: [4]

Testicles and Ovaries.
Explanted in a row
Grown by Martinowitch.
In Vitro.
Volumes at the libraries.
See how they show
Progress in Research
In Vitro
Nuns at the hospital.
Women should they know
About sexual maturation.
In Vitro

Today, scientists are able to intervene into the natural process of reproduction. Since the 1977 in vitro fertilization (IVF) that resulted in the first "test-tube baby," Louise Brown, ART have become the foundation of a thriving business. Hundreds of clinics serve women who are seeking such procedures as in vitro fertilization, gamete intrafallopian transfer, frozen embryo transfer, and other technological solutions for infertility.[5] Hundreds of advertisements in college newspapers call for egg donations from women with selected physical and behavioral traits. And there are even extreme, much-publicized practices. A virgin and a postmenopausal woman, artificially inseminated, have given birth to full-term babies. Children have been conceived with sperm extracted from dead fathers.[6]

Reproductive interventions can be costly, painful, humiliating, and dangerous to a woman's health. Yet, the demand for ART over the past decade has greatly expanded. In 1998, about 0.7% of the 3.9 million births in the United States were the result of ART. In 1998, a total of 20,143 live birth deliveries resulted from 81,899 procedures.[7] Dissenting feminists assert that these procedures give scientists and physicians more power over reproduction and perpetuate the image of women as breeders.[8] They worry about the commercialization of reproduction, observing the growing market for donor eggs and the advertising for eggs and sperm. With the development of techniques to diagnose and manipulate the genetic makeup of gametes, they foresee a new eugenics.

Motherhood as a Primal Myth

The development of new reproductive technologies has encouraged requests for ART, but the interest in these techniques also reflects the ideology of genetic essentialism that increasingly pervades popular culture perspectives on family relationships. Infertility has become a common source of desperation. Having a biologically related child has become a validation of self, a source of personal identity and continuity. Yet well before the biotechnology age, Frida Kahlo conveyed these sentiments in *Henry Ford Hospital* (1932) (Fig. 3). Surrounded by disassociated objects—a dead infant, a pelvic bone, a snail, a red violet given to her by Diego, and a defunct mechanical apparatus—Kahlo lies on a hospital bed floating in a barren industrial landscape.[9] Her life blood flows to these objects through external ribbon-like arteries, appearing in umbilical cord fashion, connecting the maker and the made. This is a motif we see repeated in several of her other paintings, as well as in works by current artists. Taken as a whole, these personally symbolic forms become the artist's narration of the devastating emotional trauma of her miscarriage. The six icons projected in space relate her ordeal to the fracture of her pelvis, which could not be fully repaired. Although she was able to conceive a child, she was also aware that, because of this injury, she could probably never carry it to full term. Yet, desperate, she risked her life on more than one occasion to try to experience motherhood.

Kahlo was conscious of the importance of family lineage and obsessed with identity. In *My Grandparents, My Parents and I* (1936) (Fig. 4), the artist represents herself as a little girl holding the ribbon of her blood line. She sees herself in this context of genealogical continuity and expresses in many of her paintings the loss, loneliness, and despair of an unfulfilled self. However, she also represents herself as the fetus in her mother's womb—a fantastic image of a longed-for, but impossible, magical daydream.[10]

Reproduction is also externally brokered through the formation of "surrogacy" contracts. In such cases, women are hired as receptacles, in effect, living test tubes for gestation. To some couples unable to conceive a child, even a vicarious pregnancy through surrogacy seems more acceptable than adopting a biological stranger. The inability to bear a biologically related child appears in popular culture as a devastating loss of the genetic bond.[11]

In both secular and religious art, childbirth has been portrayed as a miracle. The myth about the "unnatural" origins of Eve, born from Adam's rib, and the story of Mary, who conceived Jesus while she was an intact virgin, have been defined as mystical events. And in their very capacity to give birth, women are seen as both a source of life and a harbinger of the inevitability of death and decay. Contemporary artists continue to represent motherhood as an archetypal image with mythological overlays.

For Kiki Smith, the female body remains a longtime preoccupation. In *Untitled* (1989) (Fig. 5), a paper torso is suspended from the ceiling, out of which hangs a baby attached to its umbilical cord. The headless mother is a birthing vehicle, a fragmented body part. Her paper skin is permeable, osmotically open to the outside. Exposed and hollow, she evokes an eerie sense of vulnerability—of the fragile membrane between life and death, being and becoming, memory and loss.

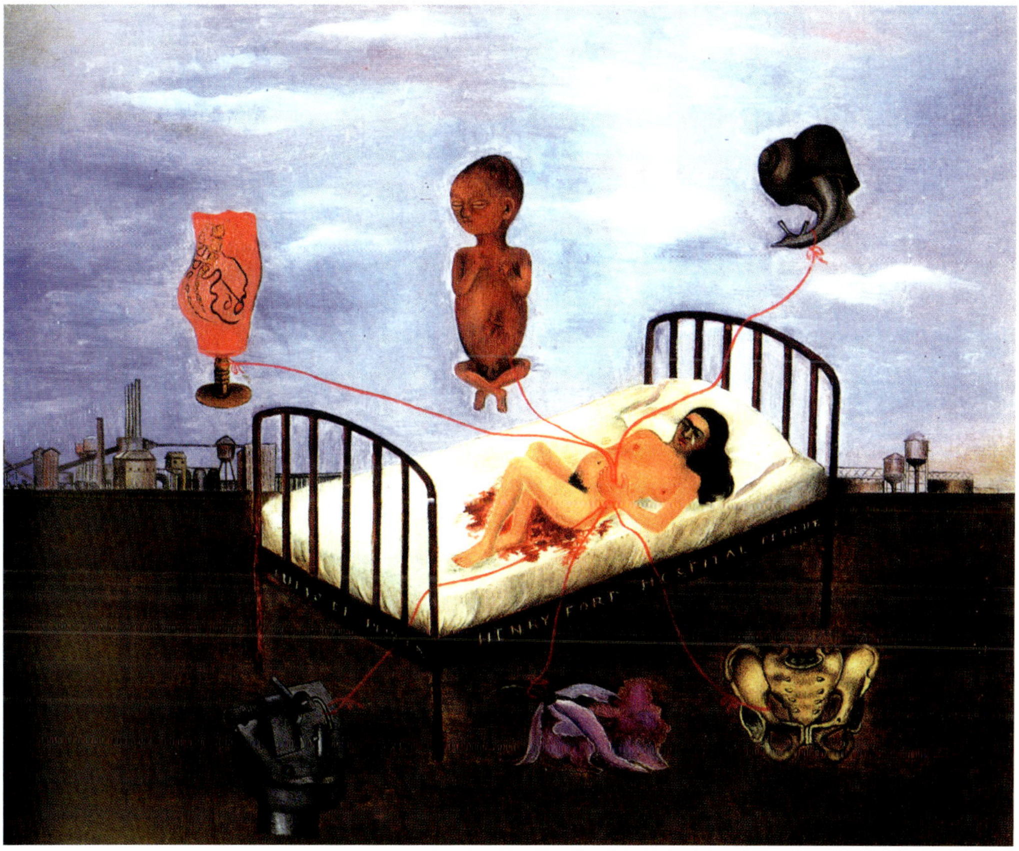

3 Frida Kahlo, *Henry Ford Hospital*, 1932.

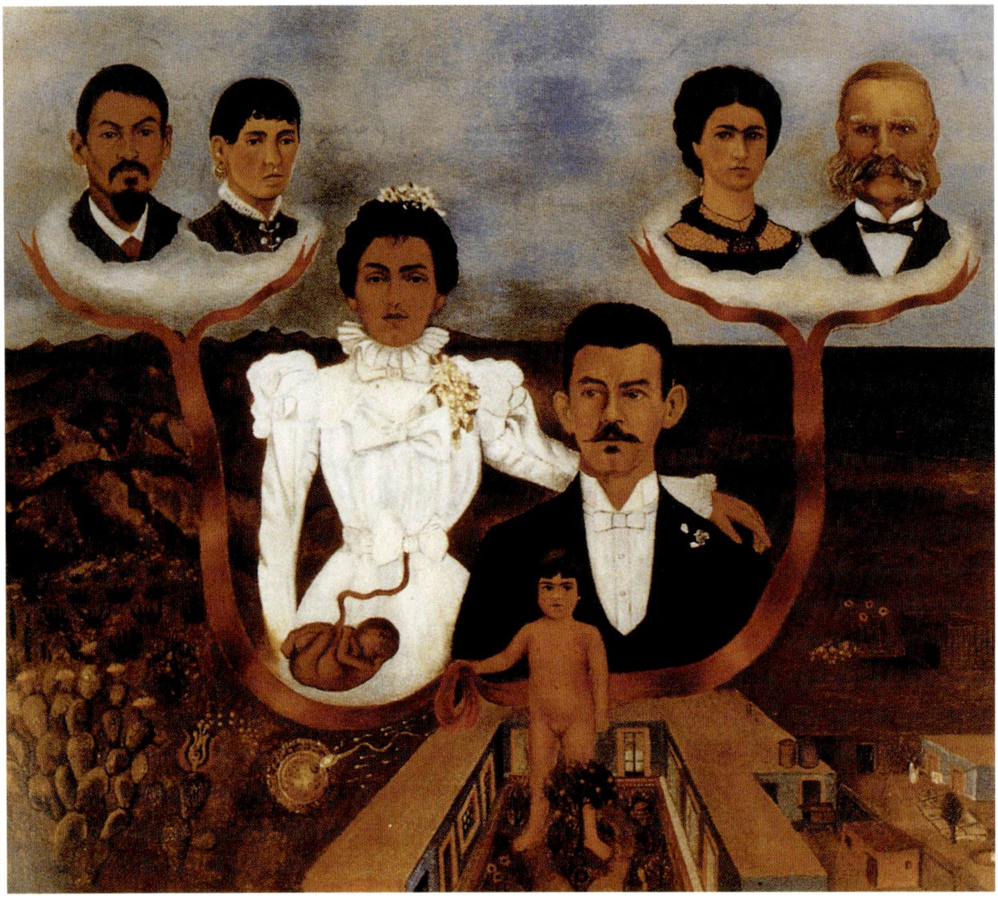

4 Frida Kahlo, *My Grandparents, My Parents and I,* 1936.

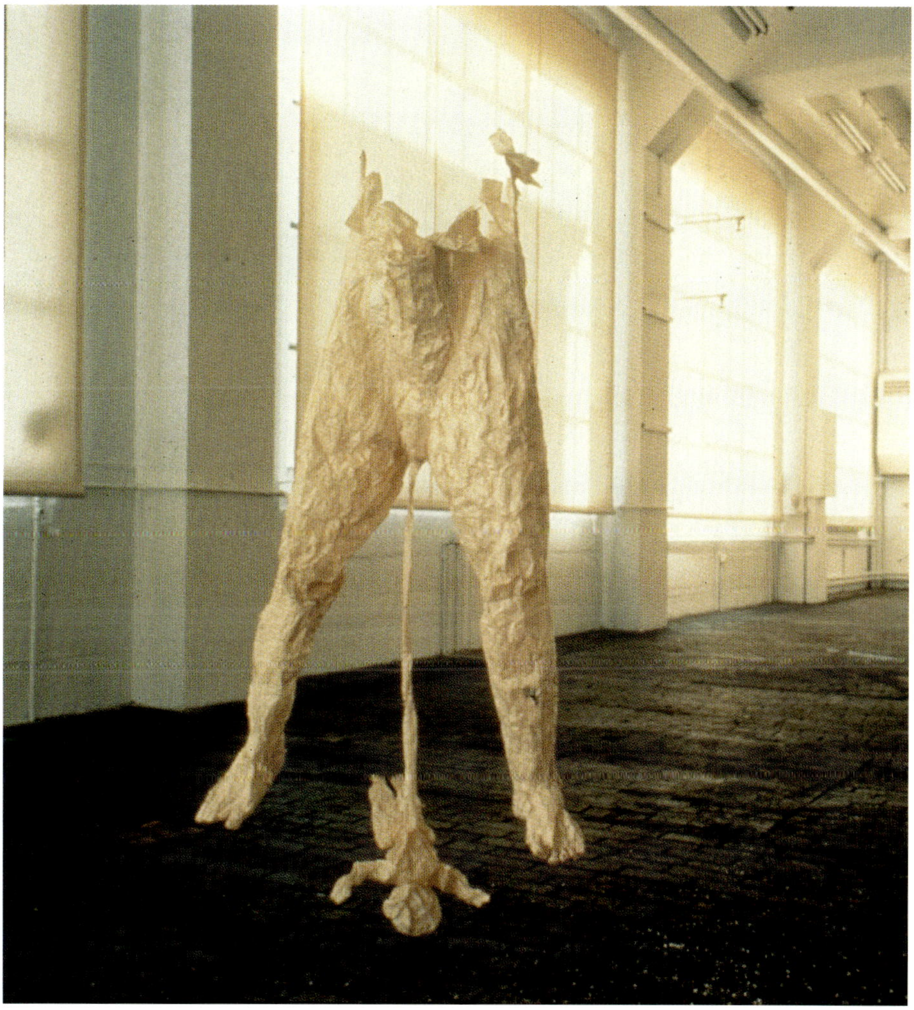

5 Kiki Smith, *Untitled*, 1989.

A penetrating version appears in Joel-Peter Witkin's *Portrait of the Holocaust* (1982) (Fig. 6), based on the mythological archetype of the "Terrible Mother." Carrying over his affinity for the extreme, Witkin's masked mother is a monstrous breeding vessel, performing a gruesome task by bringing babies into a world that can only foster their decay. In this deeply felt, yet disquieting, image, a hovering angel suggests that the mother's carnal offspring are innocent victims of life's ongoing process that inevitably culminates in death. The mother clutches her infants, suggesting the battling forces of Eros and Thanatos. Like Medea, she brings her children life, but also death.

Other images that speak eloquently of the aesthetics of the real highlighted by the fictional are Ariel Ruiz i Altaba's and Rosamund Purcell's "babies in bottles." In Altaba's melodramatic *The Three Graces* (2002) (Fig. 7), a mytho-

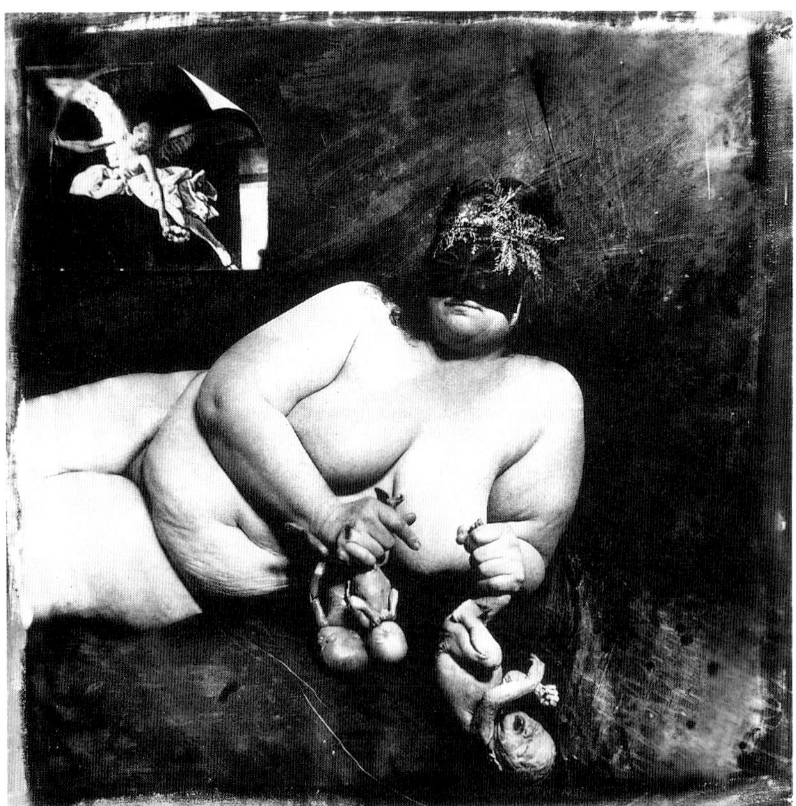

6 Joel-Peter Witkin, *Portrait of the Holocaust,* 1982.

logical subject that inspired Raphael, Rubens, and Canova is saluted again. This time the muses are invoked by an image of miscarried triplets, who find their home in Philadelphia's Mutter Museum. Fragile strings support the posture of these beings who are pictured delicately as a legend in life's course. Neither mortal (they were never born) nor godly, they hover in suspended animation. In Rosamund Purcell's Lilliputian-like specimen, a small glass-encased fetus from the Universiteitmuseum in Utrecht (Fig. 8) reflects society's fetishtic enrapture with the parallels between womb and tomb.

The fragility of life is also the theme of Marc Quinn's *Lucas* (2001) (Fig. 9), an atypical portrait of his three-day-old son. Here the mother is not visibly present, but her presence is inferred by the use of her placenta. Lucas' umbilical cord, along with his mother's placenta, has been pulverized and incorporated into the sculptural material. The infant in Quinn's delicate

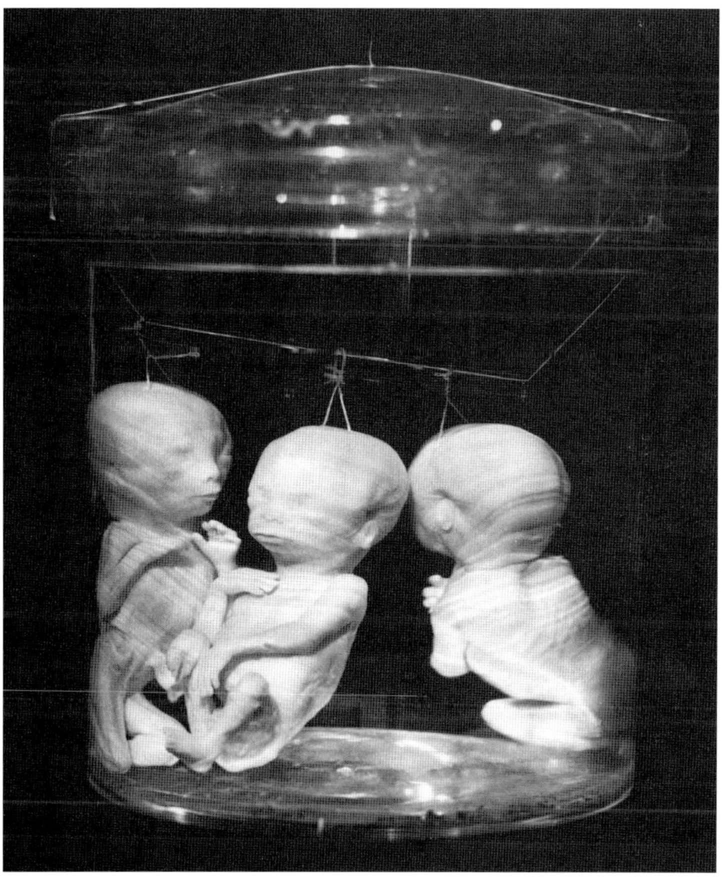

7 Ariel Ruiz i Altaba, *The Three Graces,* 2002.

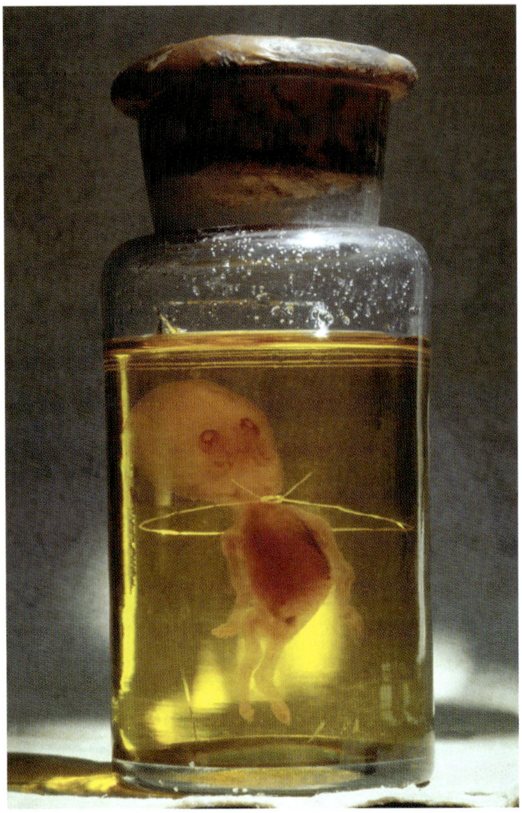

8 Rosamund Purcell, *Stained and Cleared Specimen of Hydrocephalic Child (Tiny Fetus),* 1995, from the Universiteitmuseum, Utrecht.

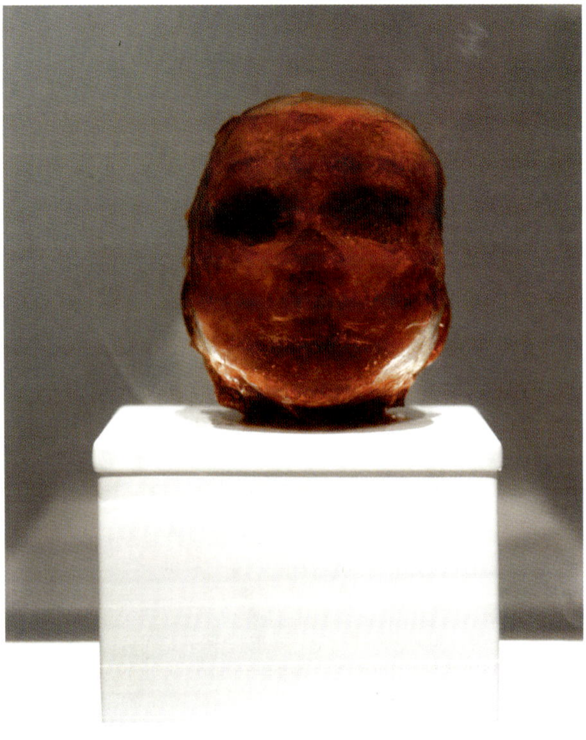

9 Marc Quinn, *Lucas,* 2001.

sculpture is equipped with this birth matter frozen in time with its own refrigeration unit. It appears like a religious shrine, an ode to the evidence of a miracle.

The Status of the Unborn

Fetuses themselves have assumed a new status as they are transformed through imaging technologies into persons.[12] One quite remarkable aspect of the new reproductive technologies is the growing importance of visual representations through the use of sonograms. Sonograms, images produced by sound waves that register on a computer screen, provide precise in utero views of the fetus and placenta.[13] They are ultrasound images whose clarity has been greatly enhanced since their initial use in prenatal imaging (Fig. 10).

Swedish photographer Lennart Nilsson first exposed the once opaque living womb to public view in 1965 with a series of fetal photographs. In a *Life* magazine article entitled "Drama of Life Before Birth,"[14] his high-resolution images captured the fetal heartbeat and face. His photographs connected Huxley's fantasy of the ectogenic baby to its biological equivalent—the fetus encased in its transparent embryonic sac (Fig. 11).

Since Nilsson, the fetus has become the subject of astonishing medical photographs of fetal development.[15] Using three-dimensional computer and scanning programs, Alexander Tsiaras isolates the different biological systems that comprise the body of the fetus—its circulatory, skeletal, and nervous systems (Fig. 12a,b). By separating out the body's organization, Tsiaras remodels the fetus into an array of simultaneously operating biological tracts. They are at once wonderfully bizarre and organically floral. In other work, such as "12 week old fetus" (Fig. 12b), the artist-scientist transfigures a fetus into a pink doll-like baby.

Alternatively, commenting on the mysterious forces of the intrauterine experience, photographer Adam Fuss imaginatively transforms that concealed reality into an external vision. By placing a baby in water along with photographic paper, he is able to use the child directly as a photo-template. The ripple marks made by the baby floating on the surface of the water visually simulate intrauterine life. Although *Invocation* (1992) (Fig. 13) is a silhouette of a full-term infant, this timeless image appears as a fetus in a placental sac.[16] Today, prospective parents send sonographic images of their fetuses to friends and families as their very first baby pictures. Such images of the fetus have also turned into a political tool. Anti-abortionists disseminate pictures of fetuses in videos such as *The Silent Scream* and *S'Aline's Solution* to portray life from the perspective of the fetus and to buttress their claim that life begins not at birth but at conception. Such images have changed the terms in which reproduction—and abortion—are framed, understood, and debated.[17]

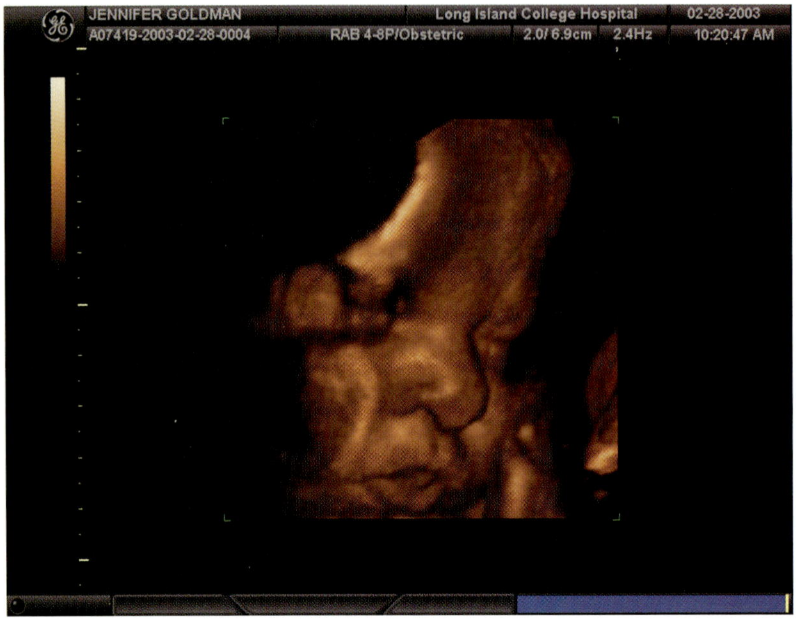

10 Dr. Andrew Goldman, Sonogram of 38-week fetus, now born and named Phineas Goldman.

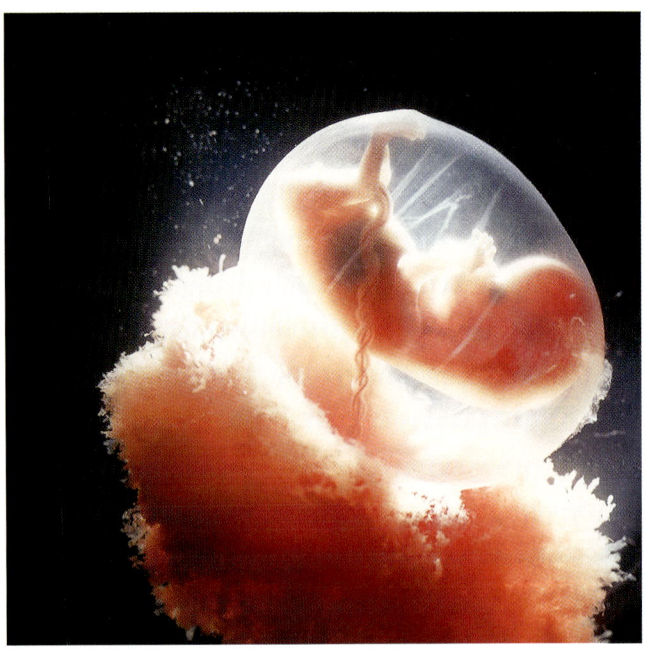

11 Lennart Nilsson, photograph, 1965.

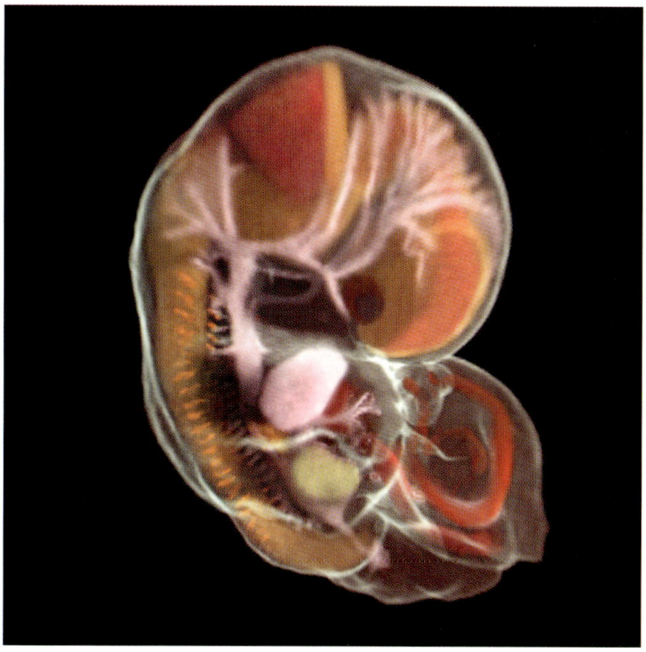

12a Alexander Tsiaras, 3-D image.

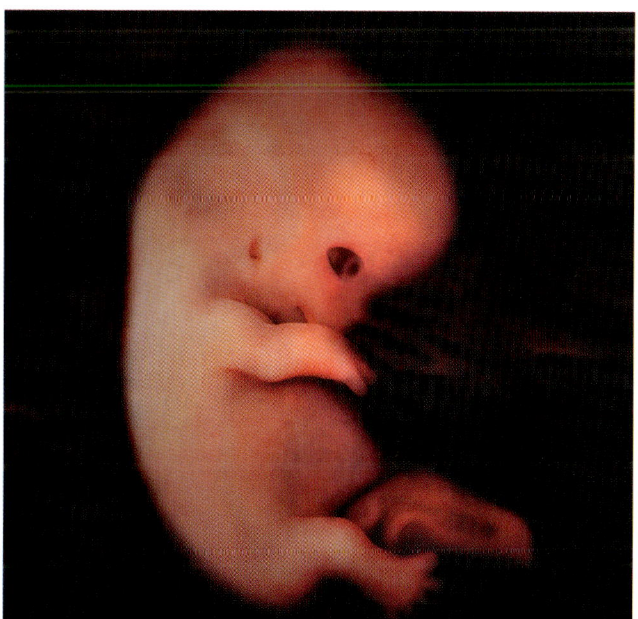

12b Alexander Tsiaras, 3-D image.

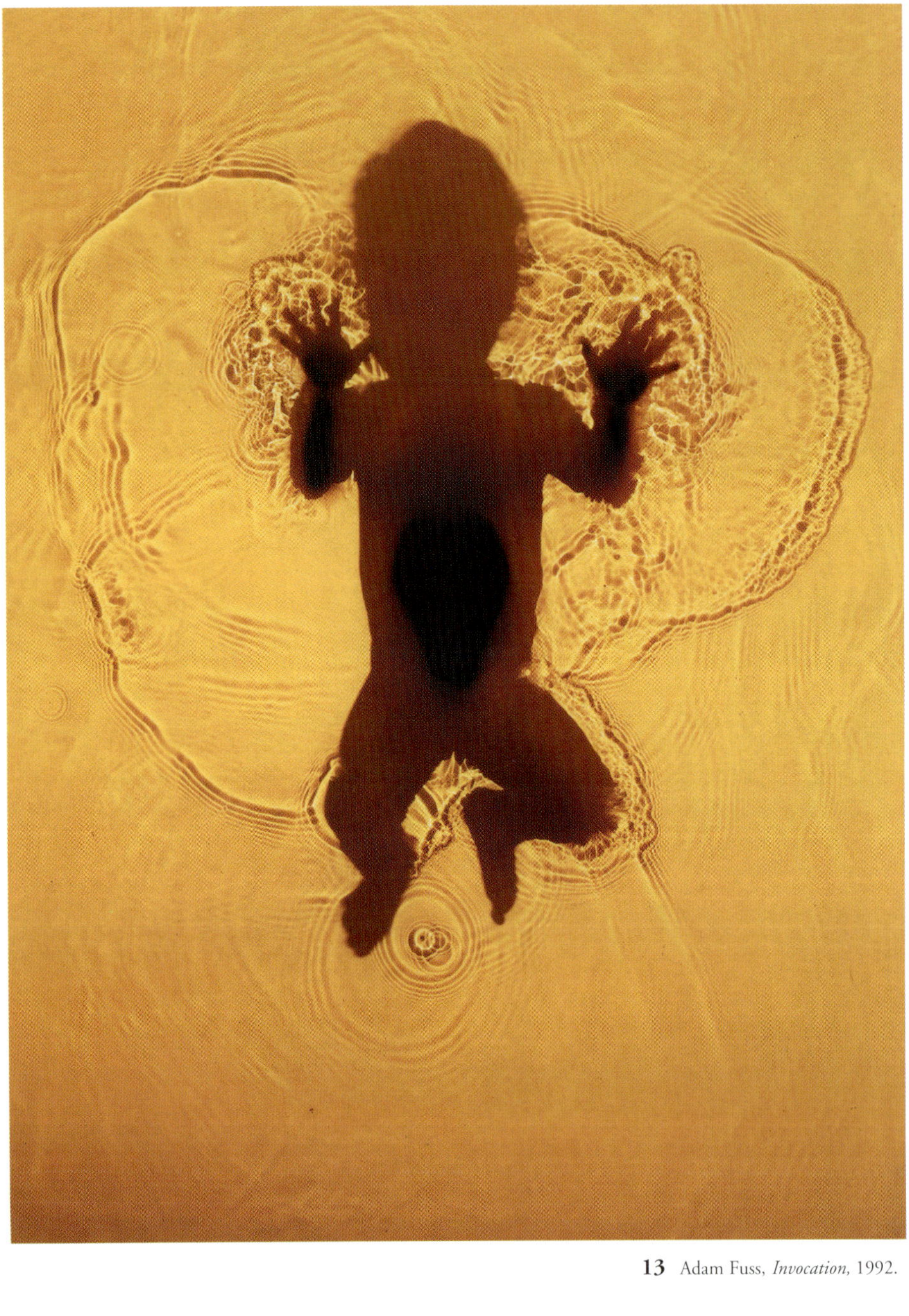

13 Adam Fuss, *Invocation,* 1992.

There is also a darker side of these images as the fetus becomes defined as less a mystery than material. At a Shanghai arts festival in 2001, Chinese performance artist, Zhu Lu, washed a 6-month-old dead fetus from an abortion and put its dismembered parts in his mouth. His performance was filmed on television and commentators speculated that the artist was immortalizing, through art, infants with the "infantile disease rampant in China known as being a baby girl."[18]

New reproductive technologies and the images used to implement them have redefined reproduction as an artificial act that is open to a startling degree of intervention, manipulation, and scientific control.[19] The possibility of preimplantation diagnosis, intended for screening embryos in order to avoid serious inherited disorders, can be extended to the screening of embryos for behavioral traits such as intelligence, sexual orientation, or desired cosmetic characteristics. Such possibilities have generated futuristic fantasies, often encouraged by scientists who, in their popular writings, advocate control of our evolutionary destiny. Lee Silver, for example, projects the possibilities of interventions that would control physical characteristics and mental capacities.[20] Gregory Stock sees "redesigning humans [as] our inevitable genetic future...the ultimate expression of our humanity.[21] A Johns Hopkins University survey found that 20% of the population approve of genetic engineering to create "a better quality child."[22] Inevitably, reproductive interventions are associated with eugenics—the breeding of "better babies."

Cloning

The term "cloning" was originally applied to botanical specimens, created through a technique of asexual reproduction in which part of the parent stock in the form of a "cutting" was used to propagate offspring. Among the earliest nuclear transfer experiments were those conducted by Robert Briggs and Thomas King in 1952, and by John Gordon, who experimented with cloning frogs in the early 1960s. Following experiments on the manipulation of the reproductive process during the 1960s, the term "cloning" became associated with human biological engineering. Through a technique of making multiple copies of a segment of DNA, genetic information from a single cell can program the development of new individuals, and implies the possibility of creating animals—and people—by design. Cloning has hence become the most controversial of all reproductive technologies.[23]

In 1993, biologists at George Washington University "twinned" a nonviable human embryo. The media covered the experiment as if it had yielded a technology for the mass production of human beings. Whereas the scientists viewed the research as a means for helping infertile patients, journalists envisioned selective breeding factories, breeding of children as organ donors, and a cloning industry for selling multiples of human beings. These were the images that shaped the subsequent public response to the cloning of a lamb at the Roslin Institute in Edinburgh in February 1997.

Dolly was produced from the genetic material in the frozen udder cells of a 6-year-old sheep. This cloning of a higher mammal from an adult cell has been described as one of the most important scientific accomplishments of the 20th century, comparable to the splitting of the atom or the discovery of the double helix.[24] Dolly became a defining icon for a new age of reproduction, but she was also a kind of Rorschach test.[25] The public response to the (re)production of a lamb by cloning a cultured cell line reflected apocalyptic fears in both religious and secular communities that since have more broadly confronted genetic engineering. Dolly offered opportunities for rational management of bodies—the complete genetic control of cows, sheep, and humans as well. She offered a specter of reproductive decisions that turn bodies into intentional products, manufactured by design. Chris Buck's chromogenic portrait of *Dolly* (1997) (Fig. 14), however, presents her sentient nature to us. Contrasted against a dark background, Dolly's textured fur and fluid eyes make her feel pet-like, an animal we are comfortable with.[26]

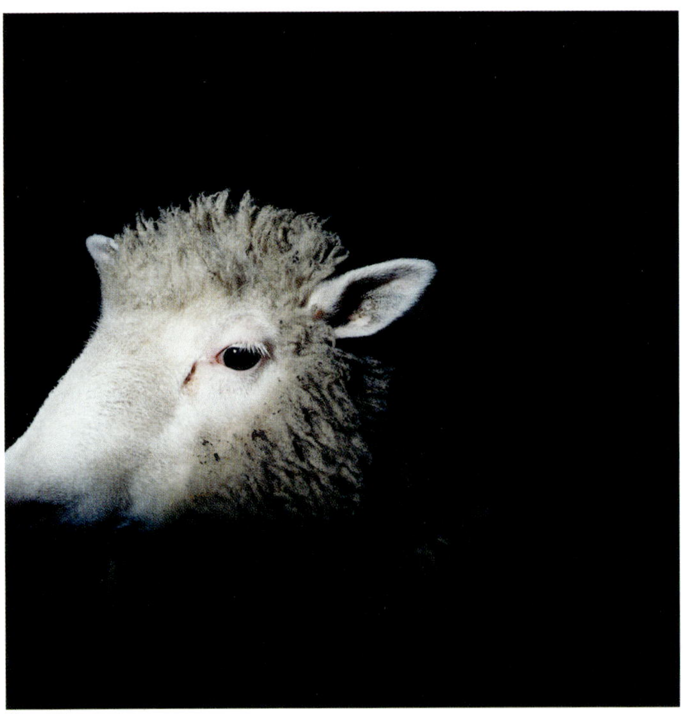

14 Chris Buck, *Dolly*, 1997.

The specter seemed to materialize at the end of 2002 when three groups of fertility scientists announced they were overseeing the births of cloned humans. One of these, Clonaid, a company organized by a religious sect, the Raelians, announced that they had cloned a baby—not surprisingly—called Eve. Although the incident was an apparent hoax, it spawned, not a clone, but a crush of media and philosophical inquiry about the ethics of cloning procedures.

Contemporary artists have addressed the issues raised by cloning. Performance artist Laurie Anderson, in a videotape *What You Mean We?* (1986) (Fig. 15), created a clone of herself. The scene opens with Anderson, on a black stage, being interviewed. She talks about how busy she has become—records, books, films, touring the world. With all these commitments, she doesn't have time to do her work any more. Cloning, although still in its early stages, seemed like a solution: she could set the standard for the attributes she required: duplicating speech patterns, musical ability, and logic. But her clone was hardly an identical Laurie. Whereas the performer appears tall and slender with an impish, harmonious voice, her clone was male, short, with a large head and deep voice. Later, we see the clone in an interview redux, looking much like the one we first saw with Laurie. Now, he says that he is so busy he has decided to produce his own clone. The camera pans out to view the clone's clone—a tall female figure. This second-generation clone again is not an identical image, but rather has a small head, red hair, and a long and slender body.[27] The tape ends with the clone's clone alone, looking uncomfortable and saying nothing.

Dolly also appears in act 3 of *Three Tales* (2002), an opera by musician Steve Reich and video artist Beryl Korot. This multimedia performance dramatizes three events as symbolic of the double-edged nature of modern technology. The first is the explosion of the Hindenburg, the second refers to the atomic tests in the Bikini Islands, and the third explores concerns about cloning and artificial intelligence. The piece is performed in full regalia with a 30-foot projection screen, ten musicians, a conductor, two sopranos, and three tenors. The collaborative team uses orchestrated sounds, narrative expression, newsreel documentation, and repeated graphic images to create an extravaganza. In one frame, a pipette extracts the nucleus from a cell (Fig. 16). Well-known scientists including James Watson, Richard Dawkins, and Stephen J. Gould are pictured as "talking heads" throughout the tale. References to Adam and the Tree of Knowledge support this operatic meditation on the role of science in society. Dolly becomes a vehicle through which to address the implications of genetic engineering and the dangers of premature decisions about transformative technologies.[28]

Facts and fictions cross in a world of cloning as the distinction between fantasy and reality is blurred. From the 1978 film *Boys of Brazil* to the 2001 Star Wars film, *Attack of the Clones,* screenwriters and directors have featured

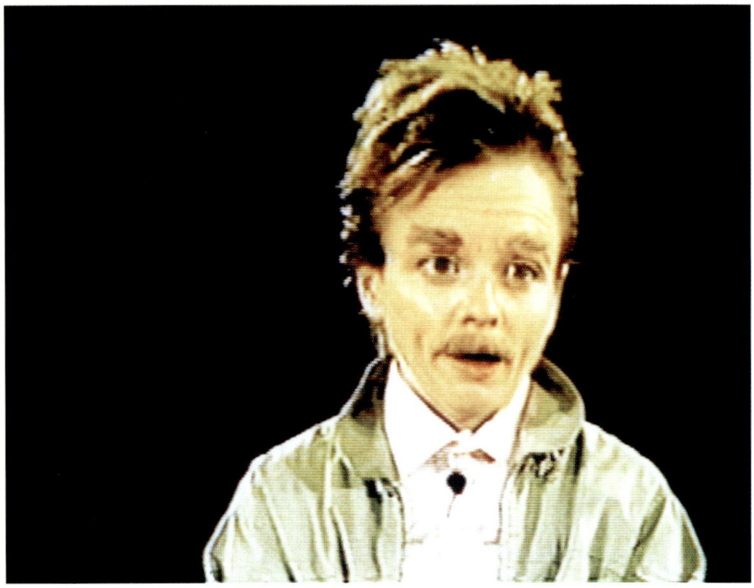

15 Laurie Anderson, *What You Mean We?* (scan of video still), 1986.

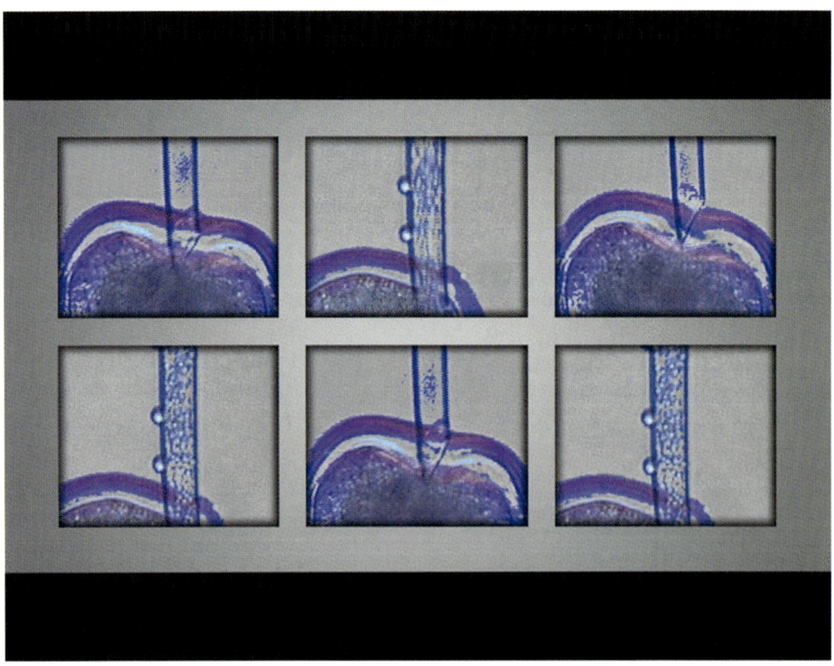

16 Beryl Korot and Steve Reich, from *Three Tales* (Act 3 *Dolly*) (video still), 2002. Image of a pipette extracting nucleus of cell.

cloning as a science fiction nightmare. In the 1999 science fiction film *Matrix,* babies are cloned and farmed to be used as an energy source, a living battery that would generate heat for the machines that have taken over the world.

Inventors and investors saw the potential for commercial cloning ventures by patenting the techniques for creating Dolly. Anthropologist Sarah Franklin described the cloning of Dolly as the creation of "breedwealth," a new kind of genetic capital.[29] For Dolly was both born and bred—a mixture of nature, culture, and corporate hubris. By moving the technology of cloning out of the laboratory and into the public spotlight, she created a remarkable mix of awe and anxiety that has come to focus on cloning's eugenic implications.

Appearing to promise wondrous new command over nature as well as terrifying dehumanization, cloning has unequivocally gripped the popular imagination. Cloning has been woven into beliefs about family, personal identity, the nature of authenticity, and the authenticity of nature. Above all, many fear that this and other reproductive technologies, combined with growing possibilities of genetic manipulation, are the harbinger of a new eugenics. If cloning holds a promise of creating perfect sheep, cows, pigs, and chickens, why not perfect people?[30] Why not use cloning to produce and reproduce perfect babies? Why not "seize control of our evolutionary future?"[31] This, after all, had once been the dream of eugenicists in the 19th and early 20th century.

Eugenics and Reproduction in the History of Art

Sir Francis Galton (1822–1911), a cousin of Charles Darwin and often called the "father of eugenics," envisioned a positive eugenics—the biological improvement of the human race by encouraging the "best people" in society to reproduce. Social Darwinist Herbert Spencer, by extension, used Darwinian concepts such as adaptation and survival of the fittest to justify a negative eugenics—letting socially problematic groups die out.[32]

Eugenicists defined the germ plasm, then believed to be hereditary material, as the essence of the person. A determinant of both physical characteristics and personality traits, "the blood will tell."[33] They claimed that this substance possessed the key to the enhancement of the human race through reproductive controls. The quality of the germ plasm would determine not only the health of the individual, but also the health of society: "The fate of the nation depends on its chromosomes," wrote Edwin Slossen, a well-known American science journalist in 1929, expressing the professional and popular view of his time.[34] The beliefs of eugenicists gained scientific credibility through studies of so-called dysgenic families such as the famous Jukes, a rural clan that over seven generations had purportedly produced 1200 bastards, beggars, prostitutes, murderers, and thieves.[35]

The excesses of the eugenics movement, culminating in the racial hygiene policies and practices of the Nazis, have had a critical influence on the way people view contemporary genetics and reproductive technologies. Recent developments in these fields have revived interest in this precursor of genetics, and there is a burgeoning literature analyzing the eugenics movement of the early 20th century and speculating on the possibilities of a new eugenics today.[36]

That the cloning of Dolly should evoke eugenic concerns is not at all surprising. Eugenicists had long applied the principles of animal breeding toward improving the human race, emphasizing the importance of "good stock." Placing moral value on eugenic fitness, the American Eugenics Society had sponsored a series of better baby and fitter family contests at state fairs throughout the country. Families with diseases or deformities, that is, "bad blood," were defined as dangerous ("born criminal") and their breeding had to be controlled.

A seldom discussed but important aspect of the eugenics movement was its organization around aesthetic premises. Eugenics was a search for "ideal beauty," or the "perfect state of nature," an idea later picked up in the Nazi vision of the Aryan ideal. Galton, who made repeated references to art in his writings, was influenced by the entrenched aesthetic tradition of his time, and, in particular, the artistic theories of Sir Joshua Reynolds, who played a seminal role in the popularization of classical aesthetics well into the 19th century.[37] For Galton, the ideal state of human perfection, defined in terms of classical aesthetics, was to be attained through selective breeding for particular physical attributes.

The eugenics movement declined in influence after 1935, its social agenda called into question by unfolding events in Germany. But its legacy hovers over new genetic programs, and atavistic eugenic thinking persists. Molecular biologist Robert Sinsheimer foresees a new eugenics based on the study of genetics. It would "permit in principle the conversion of the unfit to the highest genetic level....We should have the potential to create new genes and new qualities as yet undreamed."[38] Geneticist James Crow in 1988 described his hopes for a "compassionate eugenics" widely administered and protective of individual rights.[39] Many scientists see nothing wrong with eugenics per se, just its application through state control of reproduction.

Eugenic practices continue in countries such as Singapore and China, where the one-child family is mandatory. Chinese artist Zhang Xiaogang comments on the one-child family in his portraits of mother, father, and child. In *Bloodline: Portrait of a Family* (1993) and in *Bloodline: The Big Family, Number 2* (1995) (Fig. 17a,b), the three family members are connected with a thin red blood line, recalling the crimson ribbons in Frida Kahlo's family portraits. In Zhang Xiaogang's graphic paintings, the child in each family is a boy. The parents, as comrades, have no distinct gender characteristics aside from slight differences in their hair styles and clothing. In

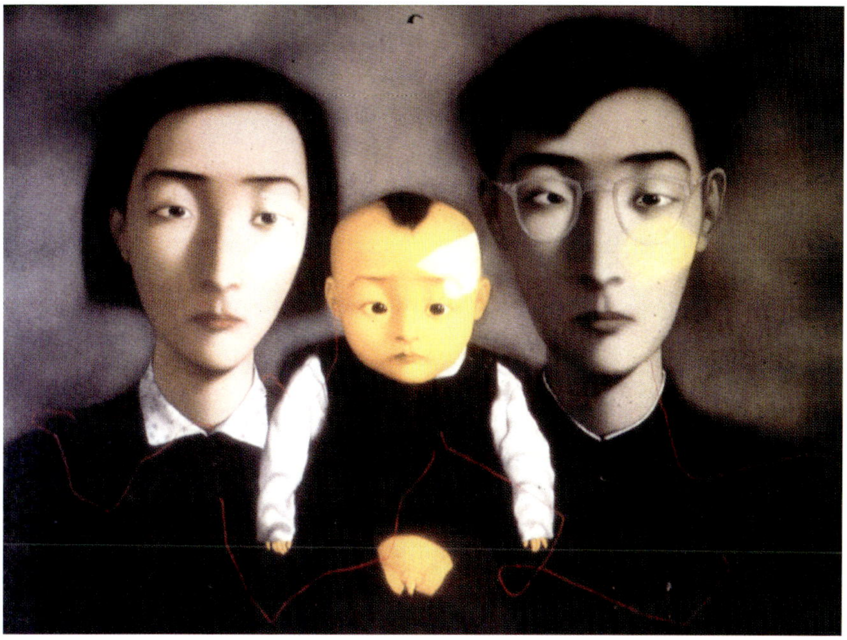

17a Zhang Xiaogang, *Bloodline: Portrait of a Family*, 1993.

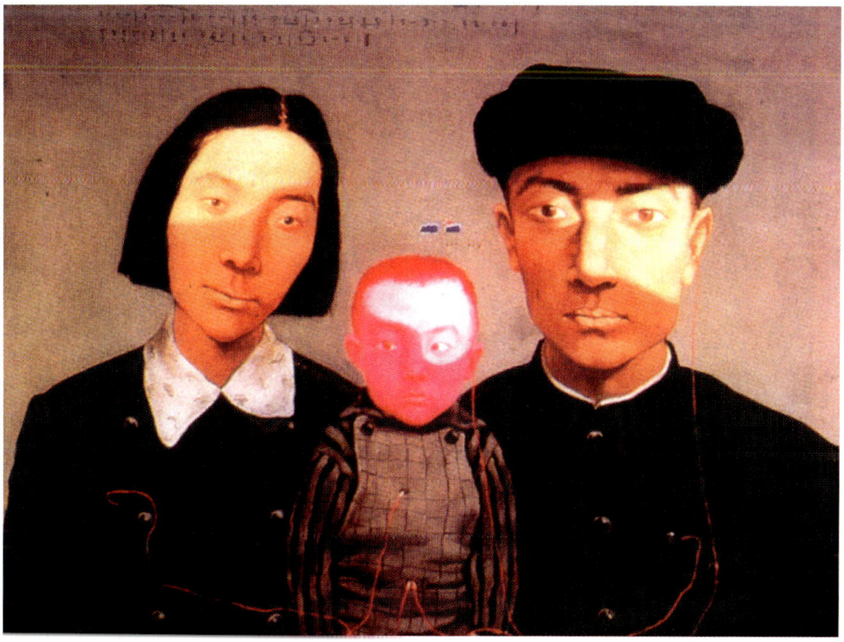

17b Zhang Xiaogang, *Bloodline: The Big Family, Number 2*, 1995.

the *Portrait of a Family*, the little boy in the center has his genitals exposed. In this rude gesture, the artist points to power preferences in families whereby the desired offspring is male, whose symbolic status represents the longevity of the family name. The portraits, marked by distorting shadows, reflect the direction of government reproductive policies. More troubling, of course, are the dwindling numbers of female children, whose whereabouts and fate are unknown.

In western societies, a new eugenics would probably not be a mandate of government policies, but a matter of individual choice. Such choices, however, would inevitably be circumscribed by economic variables and social pressures. Although a fundamental goal of the Human Genome Project is to prevent, and ultimately to eliminate, birth defects, critics remain ambivalent about the consequences of controlling human reproduction.[40]

From ART to Art: Eugenics and Reprotech in Contemporary Art

For contemporary artists, reproductive technology has become a means to explore identity, heredity, and destiny. Works of art express a flush range of dreams about science—evoking for some, euphoric fantasies, for others, the fear of science out of sway. Artists convey the questions invoked by assisted reproductive technologies in the context of broader anxieties about science and its eugenic applications. Their work dwells on several issues: experiments in animal breeding and its implications for humans, the banking of sperm, the freezing of embryos, and the significance of heredity in the genetic age.

Breeding Champions

Nicolas Rule's *Black Tie Affair* (1992) (Fig. 18) diagrams the lineage of race horses, a theme he employs in several of his paintings. Just as eugenicists used pedigree charts to analyze the inheritance of traits, so Rule uses such charts to connect well-bred successive generations of champions. Here again, as in Kahlo's and Xiaogang's paintings, the gestural lines painted in crimson simulate bloodlines—the flow of blood from one generation to another. Carefully researching the genealogy of his purebreds, Rule creates composites of family histories, tracing those traits selected as desirable or not. In the breeding of "champions," hereditary stock is carefully controlled for purity of blood.

The breeding of race horses has long relied on experience, but today scientists, hoping to make breeding a more exact science, are using genetic technologies. A UK company, Thoroughbred Genetics, will scan the chromosomal DNA of a thoroughbred for about 50 genetic markers. "We can tell how genetically healthy a horse is and its ability to stamp its progeny with similar characteristics."[41] This ability to control the quality of an animal population is called "breeding," but the same process when applied to

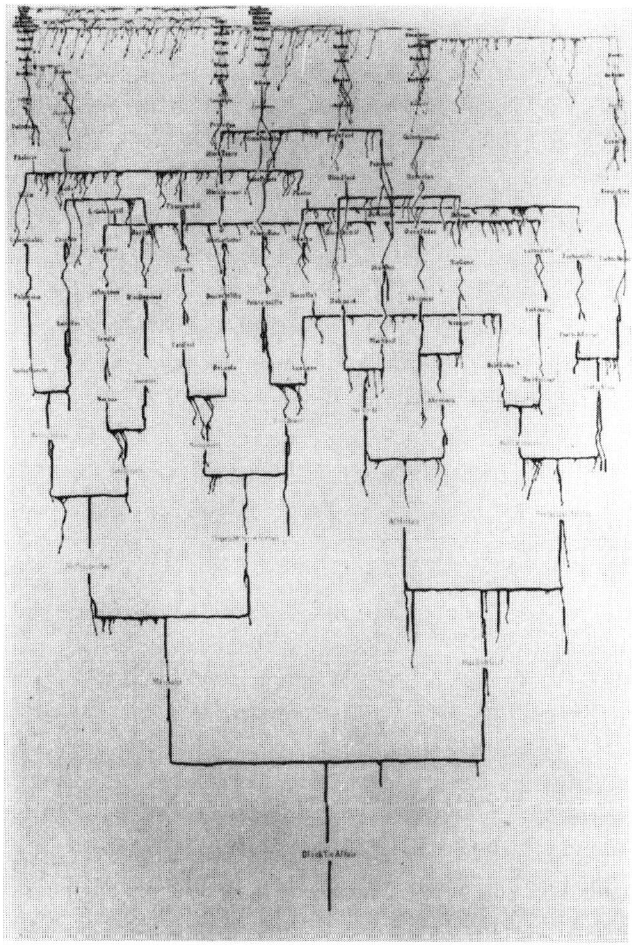

18 Nicolas Rule, *Black Tie Affair*, 1992.

humans is called "eugenics" with all its derogatory connotations recalling past abuses including involuntary sterilization.[42] The question, addressed in Rule's painting, is whether in the future, methods of selection will go beyond the breeding patterns of animals to influence human destiny.

Installation artist Andrea Zittell's interest in breeding uses actual animal specimens. *A Breeding Unit for Reassigning Flight* (1993), is an architectural sculpture in wood that serves as a habitat for her experiment in breeding chickens. The sculptural installation is constructed on ascending levels in space as little minimalist "coops." Only those chickens who can fly to a higher level in the structure are able to hatch their eggs. The social manipulation of "natural selection" for survival traits—in this case, the ability to fly high—is determined by a culturally designed environment. This work is also an elaborate visual pun on upward mobility and a cynical statement about the social biases that often enough shape evolutionary theories of genetic selection.

Breeding Humans: The Blood Will Tell

Michael Oatman's installation, *Long Shadows: Henry Perkins and the Eugenics Survey of Vermont (Vermont Pure)* (1995/2000) (Fig. 19) is part fact and part fiction—a visual essay about Henry Perkins, the first Director of the Robert Hall Fleming Museum at the University of Vermont in Burlington. Perkins, a eugenicist, was director from 1925 to 1936, and had gathered information about 4,000 Vermont residents in an effort to pass sterilization laws. Oatman's installation includes a wall of silhouettes of sterilized men and women, a visual analogy to today's "racial profiling." The profiles in the side-view portraits show physical characteristics such as small foreheads and protruding jaws, characteristics associated with the physiognomy studies of the 19th century that had inspired Degas and Daumier. In addition, genealogical charts, like those used by Galton, hang on the wall of this "reconstructed," pseudo-scientific eugenics laboratory. His purpose is to make manifest the ideology, pernicious or otherwise, underlying many scientific "facts."

Eugenics is also the focus of Christine Borland's agate mobiles in *A Treasury of Human Inheritance, Huntington's Disease, Entres' Case* (2001) (Fig. 20). In this installation, Borland had constructed a pedigree chart of five generations of a family with Huntington's disease. She suspends bits of sliced agate to create a mobile representing the generations. The variegated colors of the agate chips metaphorically refer to various symptoms that are related to this condition. Her work is derived from case studies of family disorders published in the 1930s by the Galton Institute of Eugenics, called "A Treasury of Human Inheritance."[43]

Liu Shih-Fen, in *Genetic Maps of Love* (2000), places her own blood and hair alongside illustrations of her genes encased in acrylic hemispheres. She attaches them to a wall, forming a large pink heart that juxtaposes genetic ties and love relationships. Addressing the concern that genetic testing rather than physical attraction may eventually determine matches, she expresses the fear that the "love child," that is a child that is not genetically selected, may eventually be subject to discrimination or even become extinct.[44]

Sex Cells

During the 1990s, many artists began to represent sperm, eggs, and in vitro embryos in their work. Both intimate and taboo, these artistic visualizations relate to the rapidly changing social perceptions of sex, the evolving public discourse about matters once deemed private, and the increasingly salient issues of identity politics.

Sperm have become a (sticky) subject for many artists who employ this iconography in several mediums. Kiki Smith's *Untitled (Sperm)* (1989–1990) (Fig. 21) is a cluster of 230 individual lead crystal glass sperm cells.

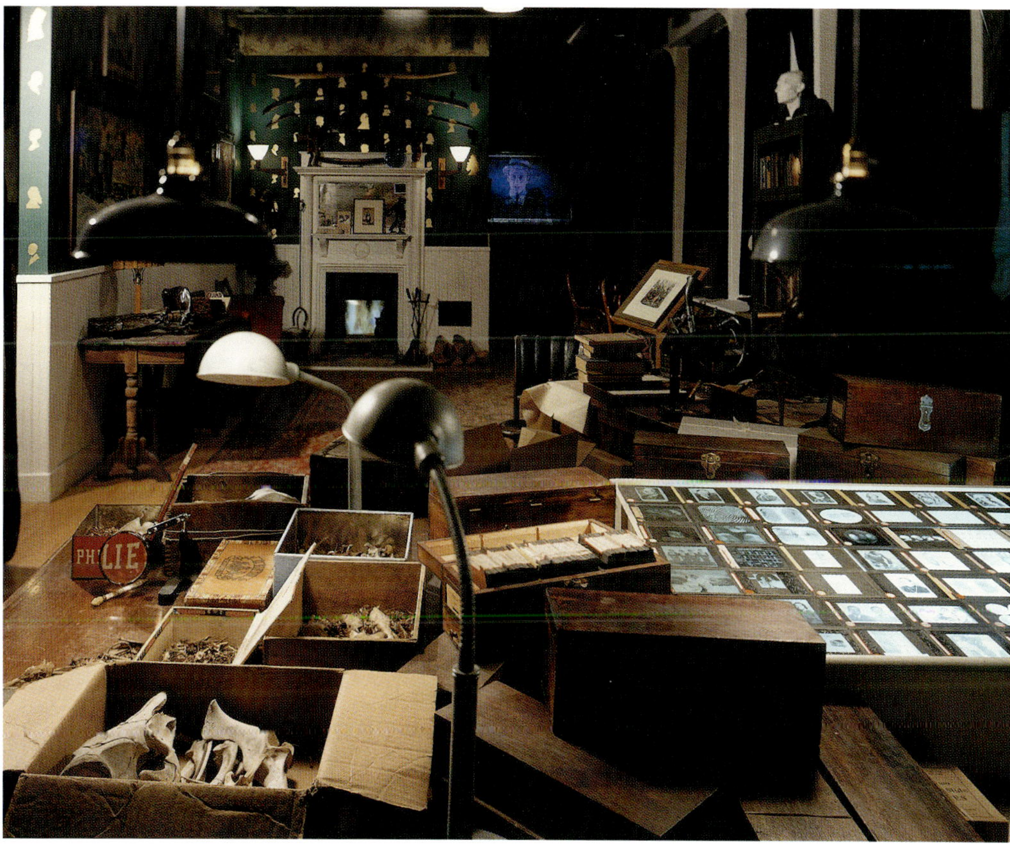

19 Michael Oatman, *Long Shadows: Henry Perkins and the Eugenics Survey of Vermont (Vermont Pure)*, 1995, recreated 2000.

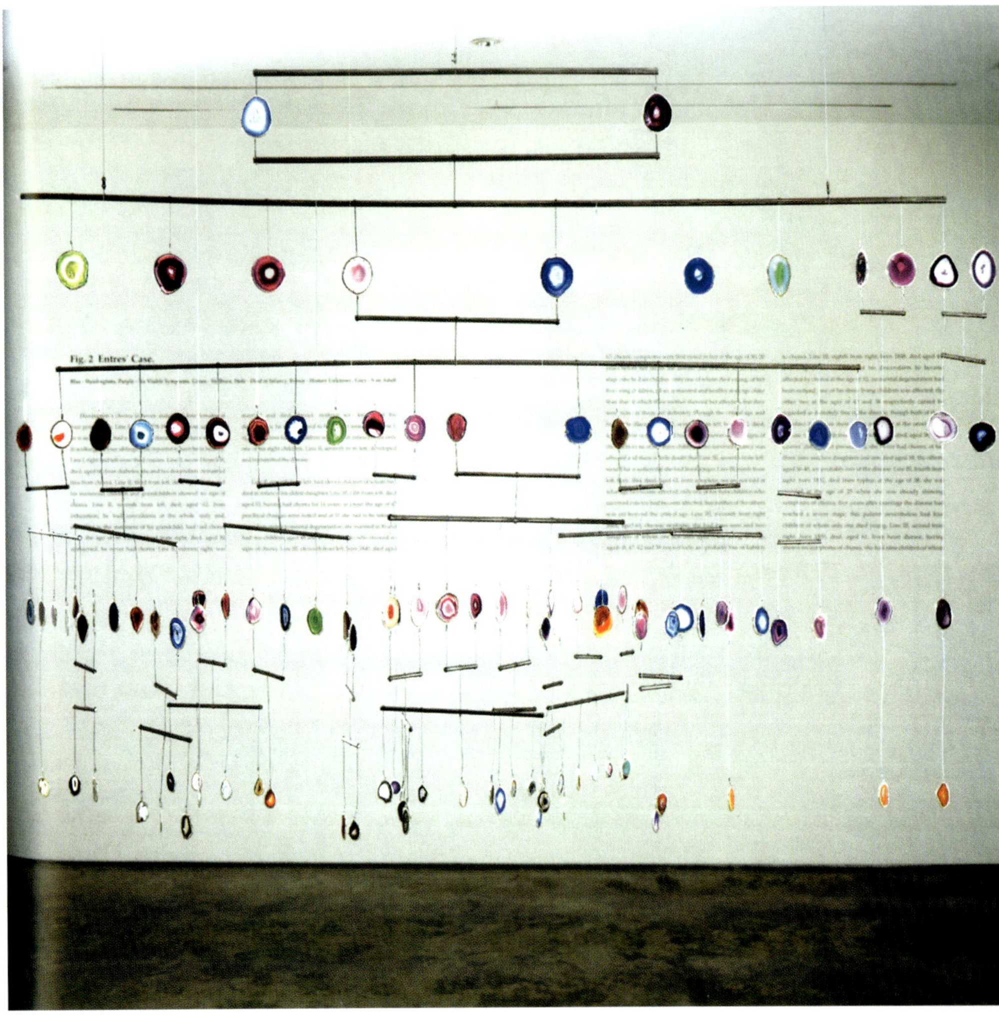

20 Christine Borland: *A Treasury of Human Inheritance, Huntington's Disease, Entres' Case*, 2001.

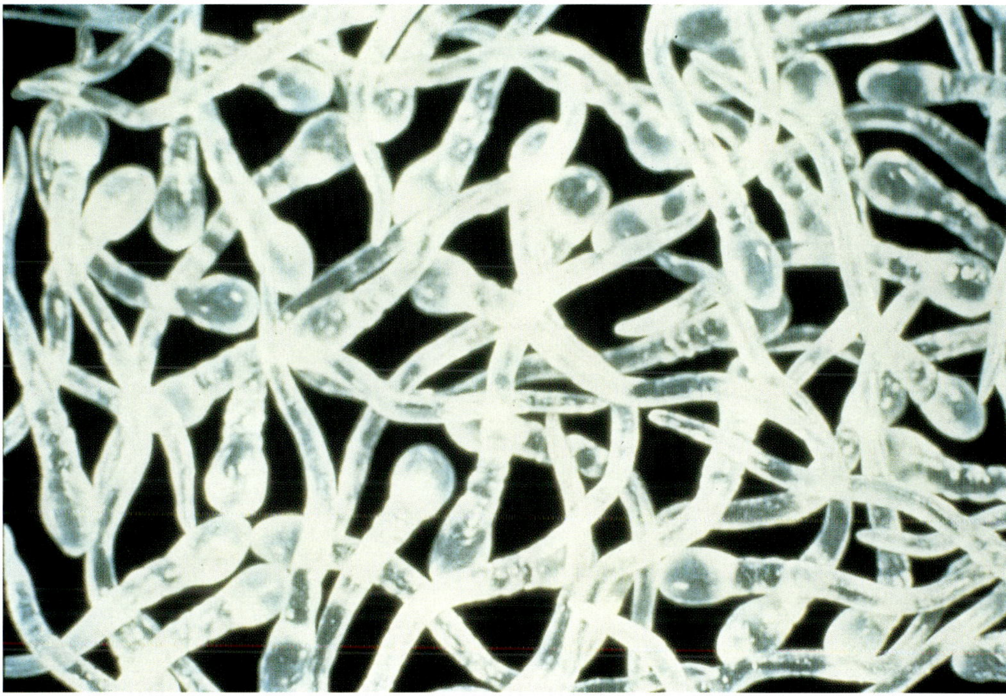

21 Kiki Smith, *Untitled* (*sperm*), Lead crystal, dimensions variable. Reproduced courtesy of the artist and Pace Wildenstein, New York, © Kiki Smith, 1989–1990.

Approximately six to eight inches in length, they spill out onto the floor as a circular scatter piece.[45] Delicate and fragile, but with a rough handmade surface, these tactile objects materialize a body fluid that is generally seen only through high-powered instruments. The sperm sculptures continue Smith's investigation into the nature of the body, its fluids, and vulnerabilities.

Andreas Serrano's photographs have become notorious—and outrageous to some —because of his use of body fluids from urine to blood to sperm as working materials in his photography tableaux. *Frozen Sperm II* (1990) (Fig. 22) captures sperm in motion as if they were shooting stars in an unmapped galaxy. As sparkling images, they visually transform this reproductive substance into a wonderous metaphysical entity.

Iñigo Manglano-Ovalle, in his installation *Banks in Pink and Blue* (1999) (Fig. 23), situates in an art gallery two large canisters of donor sperm, color-coded in pink and blue. The sperm are preserved in a cryogenic suspension with a temperature that must be maintained at –321 degrees. They can last up to 20 years.[46] The artist has made contractual agreements with the multiple donors who have contributed to the sperm bank concerning the ownership and use of the samples. The signed contracts are on the wall behind the portable repositories known as Dewar flasks. His piece, a comment on the commodification of reproduction and the growing market for sperm and eggs, touches on ethical questions: the right to market a substance that holds the characteristics of a person, the meaning of informed consent, and the way that the sperm stored in banks will ultimately be dispersed.

Sperm banking is also the subject of Kenny Schachter's *Seedbed.* This is an architectural model with shelves of test tubes that are labeled with the names of artists. It is emblazoned with the words "Artists First International Sperm Bank," a pun on the Nobel Sperm Bank founded by Frank Graham in California to encourage the reproduction of geniuses.

Embryonic Visions

Frozen embryos, a focus of much policy and legal debate, have also become a subject in art—photographed, sculpted, and drawn. How long should embryos be stored? To whom do they ultimately belong? Can surplus embryos be used in research? Is there such a thing as an "orphaned embryo?"[47]

Swiss artist Hans Danuser's *Tableau 13* from *Frozen Embryo Series I* (1998–2000) (Fig. 24) consists of abstract and minutely detailed photographs of frozen embryos that appear as a turbulent abstract field. Like fossils in an eroded rock surface, embryos—future human beings—seem to be preserved in a petrified state in the crevices between ice crystals.[48] Cryogenically suspended like a frozen food, do they have an expiration date? Under what rules and regulations do these "pre-borns" operate? Danuser's

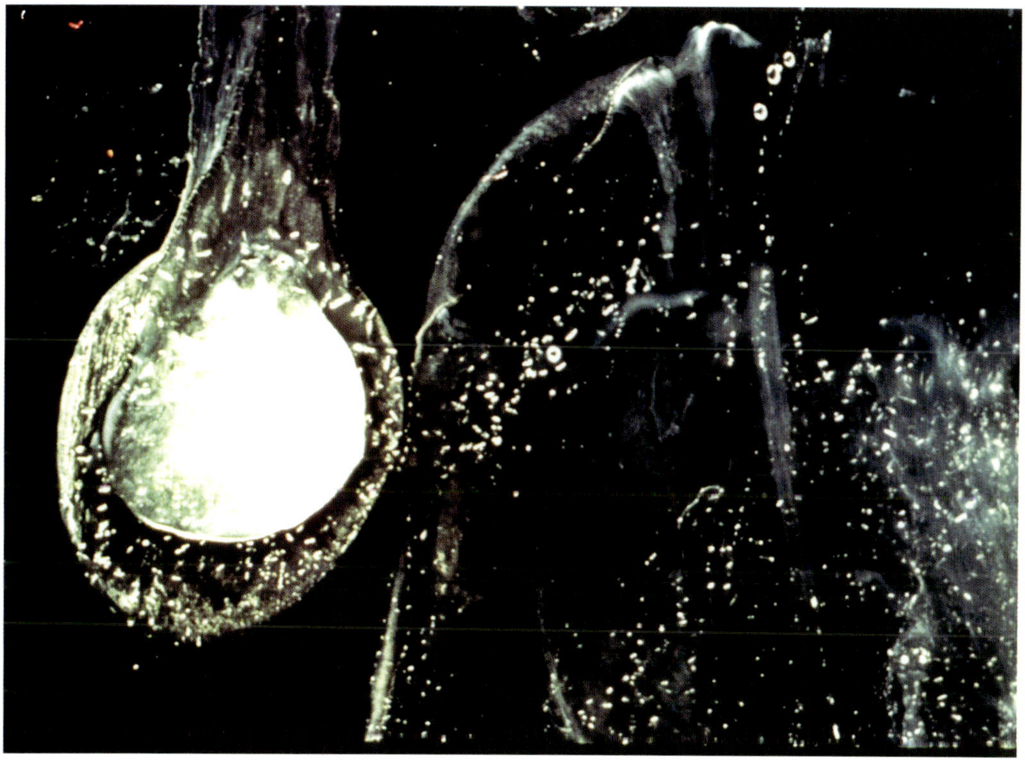

22 Andreas Serrano, *Frozen Sperm II*, 1990.

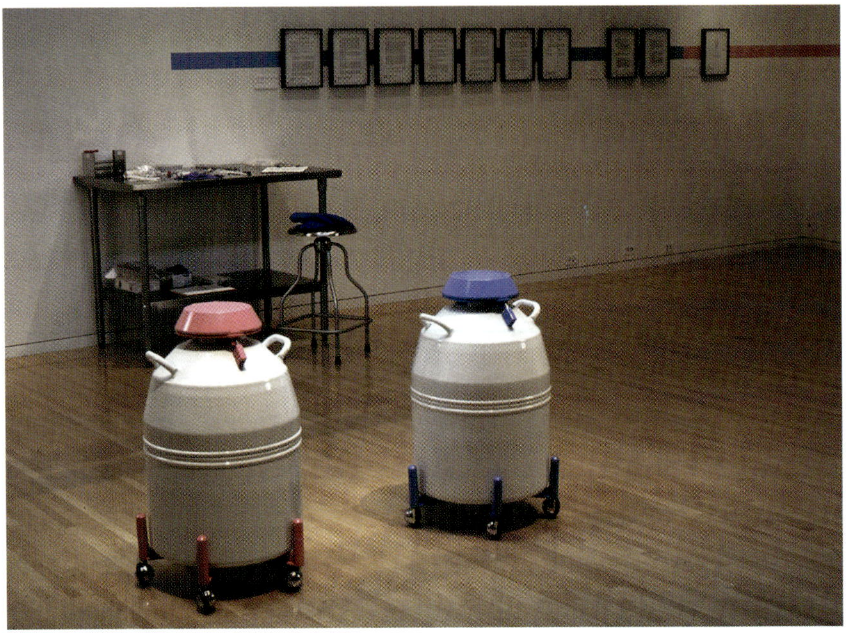

23 Iñigo Manglano-Ovalle, *Banks in Pink and Blue,* 1999.

earlier work, *Chemie II* (1989), consists of photographs of genetic engineering labs as depopulated sites with clinically distant emotional appeal. However, one image in this series contains a petri dish with two black dots. These vulnerable specks are actual embryos whose life begins in open view, but from a questionable origin and fate.[49]

Faith Wilding, a member of the collective "Critical Art Ensemble," conveys her perspectives on reproductive technology in *Embryoworlds* (1998) (Fig. 25a,b). This series of 32 watercolor drawings visualizes embryos, all painted as loose symmetrical inkblots simulating vaginas. Inscribed in each inkblot field are representational line drawings depicting aspects of assisted reproduction. Concepts include "Selection Reduction Embryo" in which one of the two embryos in a womb is being aborted. Is embryo reduction, in itself, a form of eugenics? In "Self-Surveilling Embryo," an embryo sports its own surveillance camera recording in real time the activities taking place in the womb. In this politically inspired installation Wilding explores the colonization of the female body by science.[50]

Francesco Clemente in *Silence* (2001–2002) (Fig. 26) employs watercolor to comment on mysteries of reproduction. In this subtle painting, the

24 Hans Danuser, *Tableau I3* from *Frozen Embryo Series I*, 1998–2000.

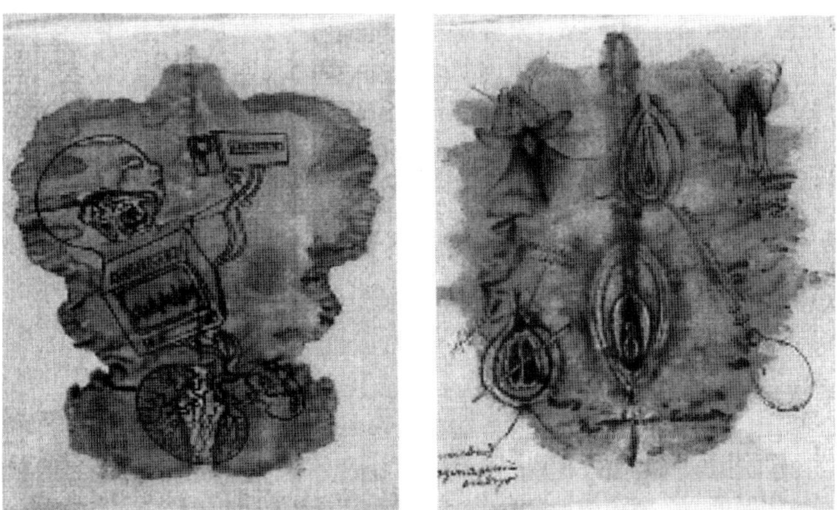

25 Faith Wilding, *Embryoworlds*, (detail), 1998.

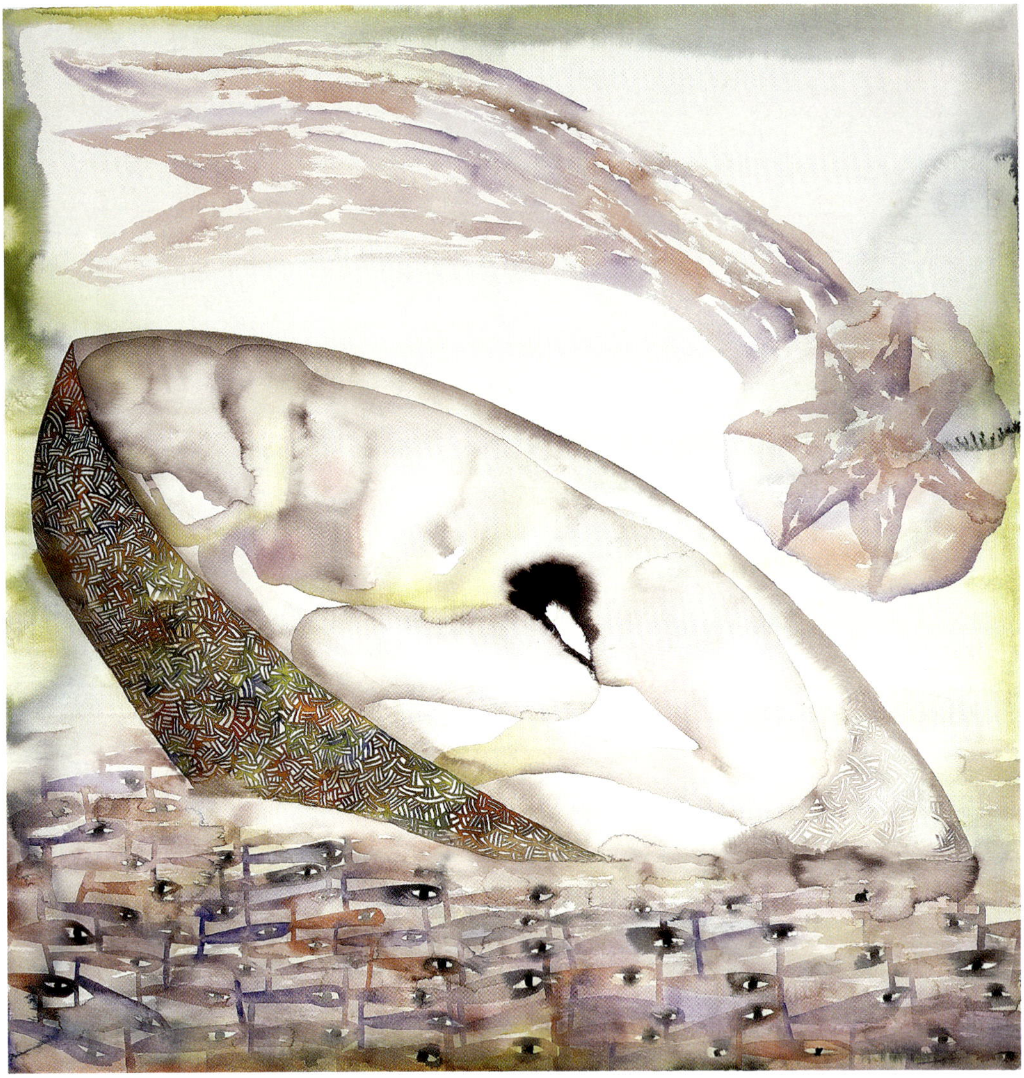

26 Francesco Clemente, *Silence,* 2001–2002.

artist converts a boat into a sexualized vaginal image that is floating over a sea of sperm-like fish. A magical star descends from the sky like a comet. This cosmological event seems to mysteriously reflect the creation of a new union.

The actual process of cell division is the subject of photographer Catherine Wagner's *Dividing Cell* (1999) (Fig. 27). Looking inside the cell, she depicts a greatly enlarged asymmetrical organic mass as it moves through the process of becoming two. In producing this photo, this artist has worked with scientists at the Weizmann Institute in Israel, and her work, relying on stark beauty for its artistic impact, is closely linked to scientific photography.

In videos and sculpture, Heather Sheehan addresses the ways in which the body and its parts can be artificially grown and nurtured through laboratory procedures. *egg* (1999) (Fig. 28) portrays an oversized egg that is nourished through a clear plastic tube attached by an external plug into its delicate surface. While incubating, the egg sits on a bed of Vaseline that looks like jelled albumen. One is led to wonder whether the egg in this vinyl environment has been artificially manipulated and, thus, to question the ethics of supporting life in the laboratory that could not exist otherwise or elsewhere.

Similar questions shape Helen Chadwick's series entitled "Unnatural Selection." In *Nebula* (1996) (Fig. 29), a series of photographs of human embryos displayed like jewels, she points to the ambiguous status of the frozen embryo. Is it a natural or artificial object? In her research at the Assisted Conception Unit at Kings College Hospital in London, Chadwick photographed dead human embryos and found their cellular patterns and partially formed features beautiful and poignant. She also discovered that doctors select embryos for implantation on the basis of aesthetic considerations and their sense of the "best" morphology as an aspect of viability. [51]

ART and Authenticity

Artists' images of ART and its association with eugenics and controlled reproduction convey concerns about the perpetuation of family, its history and blood lineage. The expression of "blood lines" is important in these images, sometimes denoted abstractly as in the red linear elements that appear in the work of Kahlo, Rule, and Zhang; and other times more literally in the visual representations of gametes as the agents of reproduction. Close to the surface in these depictions is the specter of a new eugenics.

The issue of authenticity pervading the art of ART has been a recurrent theme, especially in response to techniques of mechanical reproduction.[52] Through such techniques, Andy Warhol, in a 1963 replication of images by photo-silk screening, *Thirty Are Better Than One* (Fig. 30), in effect clones the Mona Lisa. The viewer is given not one, but numerous, images of per-

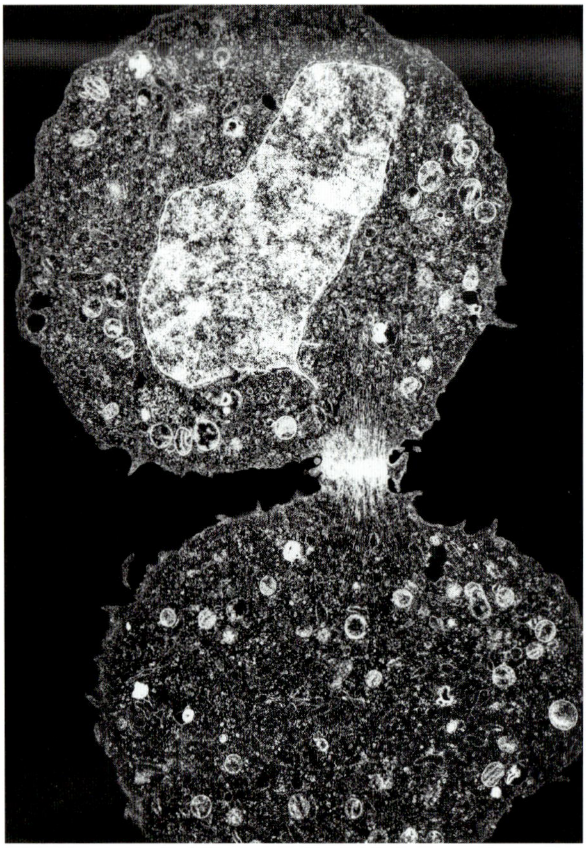

27 Catherine Wagner, *Dividing Cell*, 1999.

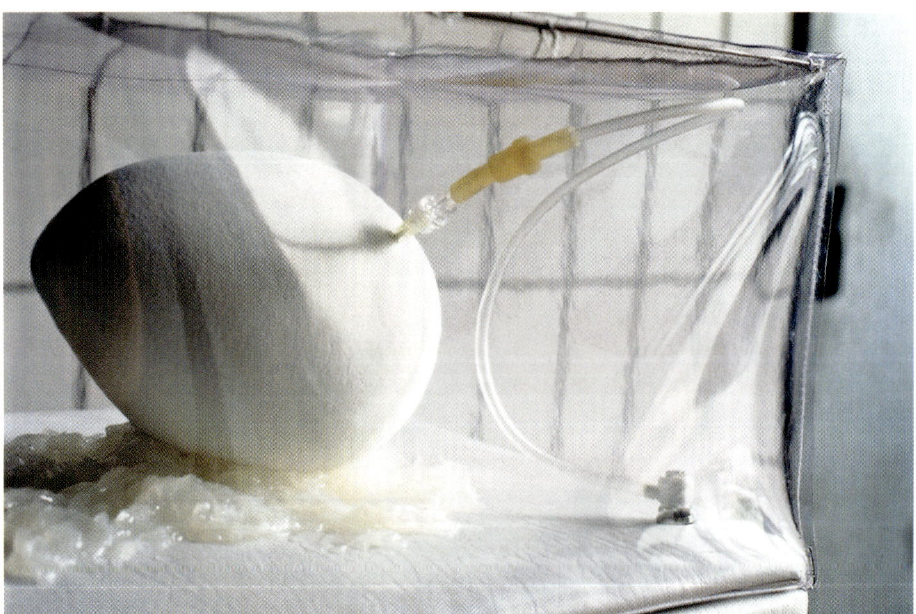

28 Heather Sheehan, *egg*, 1999.

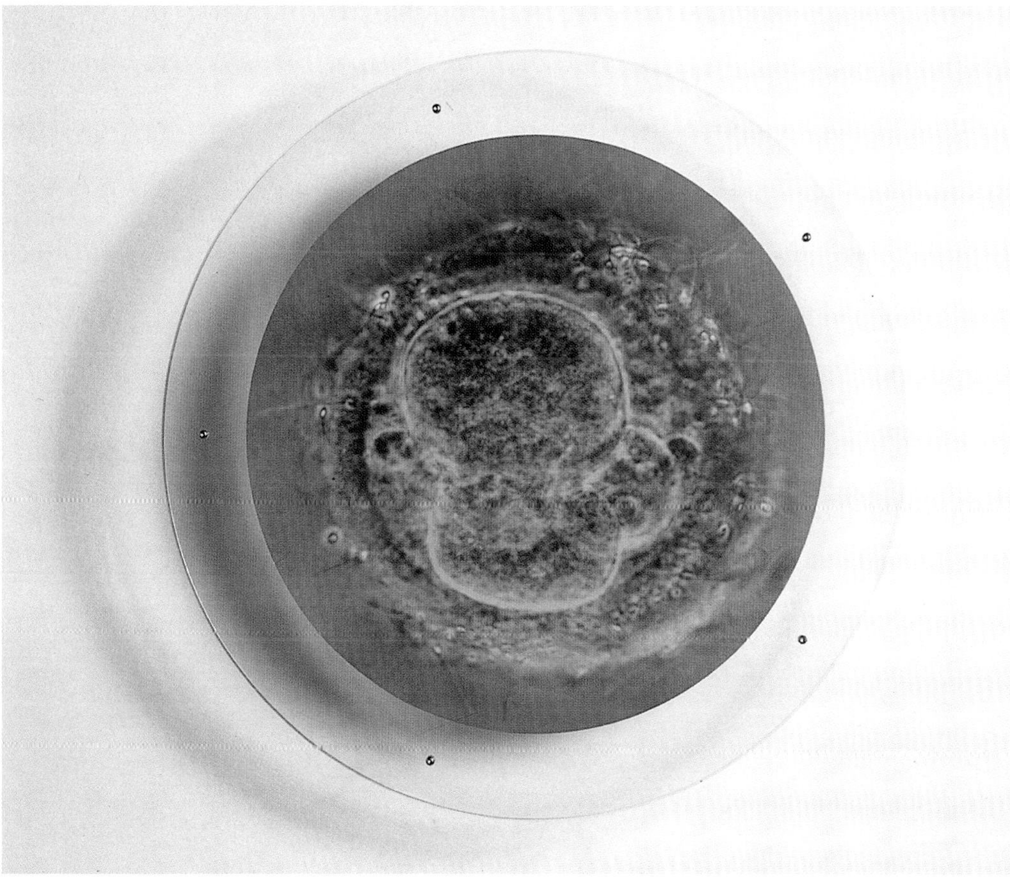

29 Helen Chadwick, *Nebula,* 1996.

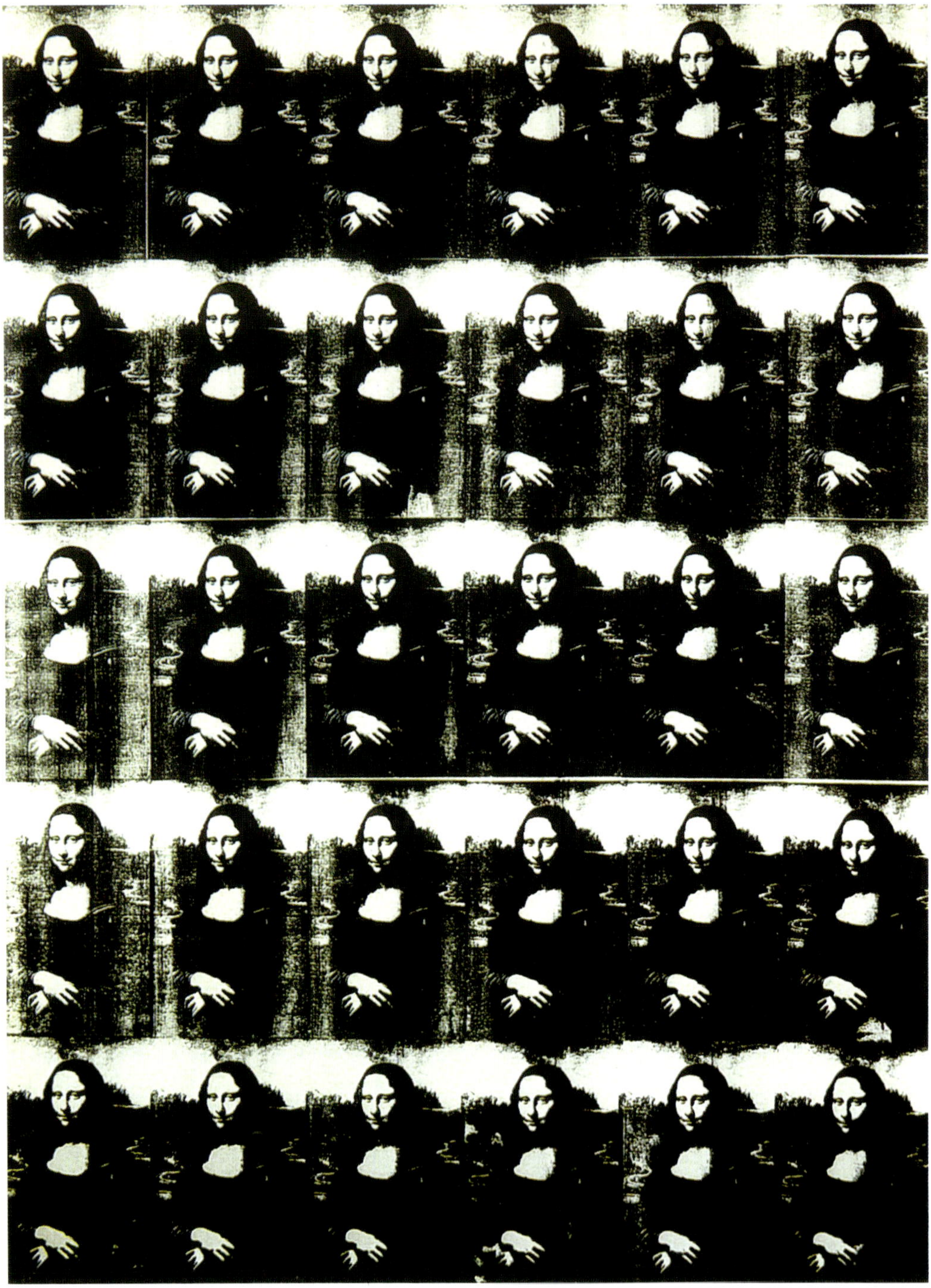

30 Andy Warhol, *Thirty Are Better Than One,* 1963.

haps the most famous painting in the world. But in the case of mechanical replication as well as in biological cloning, the sequence of identical images becomes increasingly blurred as the coded information is reprocessed. When we replicate art or clone life, what in fact do we get? How do we perceive a copy of art or a duplicate of life? What does this replication process imply with regard to the value of a unique person and personhood? Or for that matter the unique status of a work of art?

Finally, the growth of an in vitro fertilization industry has compounded concerns about exploiting women's bodies, an issue expressed by many artists from Kiki Smith to Faith Wilding. Advertisements for private fertility clinics, the banking of sperm and embryos, the calls for egg and sperm donors, and incidents of embryo theft have given the ART industry an unsavory cast by associating it with commerce. Dolly called forth visions of a cloning industry and breeding factories and elicited extravagant scenarios about persons as well as animals becoming "supermarket commodities."

Suzanne Anker's baby in the bottle evokes these concerns. It recalls the history of images that were inspired by the early experiments in ectogenesis, but also the recent cynical media descriptions of "test tube capitalists" involved in the new reproductive technologies. And it reflects a larger public concern about the commodification of science in the biotechnology age.

Notes

1. Susan Merrill Squier. *Babies in bottles: Twentieth century visions of reproductive technology.* Rutgers University Press, New Jersey, 1994. Babies in bottles appear in science fiction stories of reproduction. But they take on another meaning as artifacts of death in medical museums (Museum Vrolik in the Netherlands) and Wunderkammern collections such as Peter the Great's display of embryos in alcohol-filled jars.
2. Jon Turney. *Frankenstein's footsteps: Science, genetics and popular culture.* Yale University Press, New Haven, 1998.
3. *London Daily Express,* March 16, 1936. From Spears Archive in the history of medicine library, Wellcome Trust Archive, London, United Kingdom.
4. Ibid.
5. Valerie Hartouni. *Cultural conceptions.* University of Minnesota Press, Minneapolis, 1997.
6. Pierre Baldi. *The shattered self: The end of material evolution.* MIT Press, Cambridge, 2001, p. 46.
7. Centers For Disease Control. 2002. Use of assisted reproductive technology—United States, 1996 and 1998. *J. Am. Med. Assoc.* 287: 1521–1522.
8. Suzanne Anker. 2000. *Mothers of invention.* In S. Bee and M. Schor, ed. *Meaning: An anthology of artists writings and criticism.* Duke University Press, Durham, pp. 255–259. This concern is also expressed in fiction. Margaret Atwood, in *The handmaid's tale,* divides women into breeders and nonbreeders. The breeders are little more than slaves to reproduction.
9. Andrea Kettenmann. *Frida Kahlo 1907–1954: Pain and passion.* Taschen, Cologne, 2000.
10. Phyllis Tuchman. 2002. Frida Kahlo. *Smithsonian* November 2002: 52–60.

11. For documentation, see Chapter 4, The molecular family. In Dorothy Nelkin and Susan Lindee. *The DNA mystique: The gene as cultural icon.* W.H. Freeman, New York, 1995.
12. Debra Rosenberg. "The War Over Fetal Rights." Newsweek June 16, 2003 (cover story), pp. 40–46.
13. Rayna Rapp. 1998. Real time is prime time: The role of the sonogram in the age of mechanical reproduction. In *Cyborgs and citadels: Anthropological interventions into techno-medicine.* (ed. G. Downey et al.). University of Washington Press, Seattle.
14. See Lennart Nilsson. *A child is born.* Dell Publishing Company, New York, 1990.
15. Alexander Tsiaras. *From conception to birth.* Doubleday, New York, 2000.
16. Behind the scenes with Adam Fuss. *Art on Paper* (Sept./Oct. 2002) 7: 68–73.
17. Hartouni, op. cit, chapter 3.
18. Thomas C. Greene. 2001. Online baby muncher is an artist. *The Register* At: http://www.theregister.co.uk/content/archive/17157.html
19. The ability of science to control the natural process of birth is suggested by an experiment at the Reproductive Genetics Institute in Chicago where scientists have devised a way to create artificial sperm from any cell in a woman's body. The technique, called haploidization, involves taking half the genetic material from one cell and injecting it into another woman's egg, resulting in an embryo with the genes from both cells.
20. Lee Silver. *Remaking Eden: How genetic engineering and cloning will remake the American family.* Avon Books, New York, 1998.
21. Gregory Stock. *Redesigning humans.* Houghton Mifflin, Boston, 2002. Stock organized a genetics and culture symposium in March 2002 in which artists, designers, biologists and cultural critics were invited to discuss the issue of redesigning life.
22. Lynn Gamwell. 2003. Perceptions of science: Beyond the visible — Microscopy, nature, and art. *Science* 299: 341.
23. Cloning is a form of asexual reproduction in which an embryo is created through somatic cell nuclear transfer. The nucleus of a cell from an adult is transferred to an egg whose own nucleus has been removed. The purpose is to create a baby. This procedure differs from stem cell or therapeutic cloning, a procedure in which the embryo is created, but only to extract stem cells for medical therapy.
24. Sarah Franklin. *Dolly mixtures: Cloning, capital and immortality.* In press, 2003.
25. Dorothy Nelkin and M. Susan Lindee. 2001. Cloning in the popular imagination. In *The cloning sourcebook* (ed. A.J. Klotzko), pp. 83–94. Oxford University Press, New York.
26. Dolly is the cover image for the ICP brochure "How Human: Life in the Post-Genomic Era," curated by Carol Squiers at the International Center for Photography in NYC, February 28–June 8, 2003.
27. WGBH Educational Foundation and Canal Street Communications, 1986.
28. Julia Wolfe. 2002. Steve Reich and Beryl Korot. *Bomb Magazine.* At: http://www.bombsite.com/reich/reich.html
29. Sarah Franklin, op. cit.
30. In 2002, Ian Wilmut, the creator of Dolly, applied for a license to use nuclear transfer to create human embryos that are genetically identical to adult donor cells. The work will be in collaboration with Geron Corporation, a California firm. Gretchen Vogel. 2002. Cloning pioneer heads towards human frontier. *Science* 298: 37–39.
31. Stock, op. cit.
32. Herbert Spencer, Social Statistics (1851). See also Daniel Kevles. In *the name of eugenics.* Knopf, New York, 1985, for an analysis of Karl Pearson's papers in Chapter II.

33. Kevles, op. cit.
34. Edwin Slossen. 1929. Democracy and knowledge. In *Preface to the universe* (ed. B. Brownell). Van Nostrand, New York, 1929, p. 108.
35. Nicole Hahn Rafter. *Creating born criminals.* University of Illinois Press, Chicago, 1997, p. 38. Quoting from Richard Dugdale's *The Jukes: A study in crime, pauperism, disease, and heredity.* 1887. In 2003 new archival research turned up information indicating that Jukes family members were, in fact, neither criminals nor misfits. Some were prominent members of Ulster County society. They were an example of how scientists distorted research results for ideological reasons. Scott Christianson. 2003. Bad seed or bad science. *The New York Times* February 8.
36. See, for example, Daniel Kevles, op. cit. Dianne B. Paul. *The politics of heredity.* State University of New York Press, Albany, 1998. Elof Axel Carlson. *The unfit.* Cold Spring Harbor Laboratory Press, New York, 2001.
37. Kirby Gookin. *Eugenics and the aesthetics of ideal beauty.* Routledge, New York, in press, 2003.
38. Cited in Kevles, op. cit., p. 267.
39. James Crow. 1998. Eugenics: Must it be a dirty word? *Contemporary Psychology* 33: 9–12.
40. Dorothy Wertz. 2002. Did eugenics ever die? *Nat. Rev. Genet.* 3: 408.
41. The sport of kings becomes sport for genomics. 2002. *The Scientist* 16: 8.
42. Philip R. Reilly and Gerald R Grob. *The surgical solution: A history of involuntary sterilization in the United States.* Johns Hopkins Press, Baltimore, 1991.
43. Catherina Albano and Marina Wallace. The Genius of Genetics: A Celebration of Gregor Mendel through Science and Art, Catalog of Exhibition at Abaye of St. Thomal, Brno, Czech Republic, 2002.
44. Wellcome Trust. Working drafts: Envisioning the human genome, Two-Ten Gallery, London, January 2002.
45. Susan Tallman. 1992. Kiki Smith: Anatomy lessons. *Art in America* April pp. 147–153.
46. Marvin Heiferman and Carol Kismaret. 2001. Picturing the genetic revolution. Tang Teaching Museum, Skidmore College, New York.
47. These questions have been the focus of many lawsuits over who has rights to embryos. See, for example, York v. Jones, 717 Supp. 41 E.D. Va. 1989.
48. Hans Danuser and Urs Stahel. *Frost.* Scalo Books, Zurich, 2002.
49. James Lewis. 1990. Review of Hans Danuser exhibit at Kurt Marcus Gallery. *Art Forum* April 1990, pp. 171–172.
50. Amelia Jones. 1999. Faith Wilding and the enfleshing of painting. *N. Paradoxa* Issue 10, June.
51. Sian Ede. 2002. Science and the contemporary visual arts. *Public Understanding of Science* 11: 66.
52. For a classic discussion, see Walter Benjamin. 1968. Art in the age of mechanical reproduction. In *Illuminations* (ed. H. Arendt), pp. 217–253. Schoken Books, New York.

Commodification: Genes for Sale

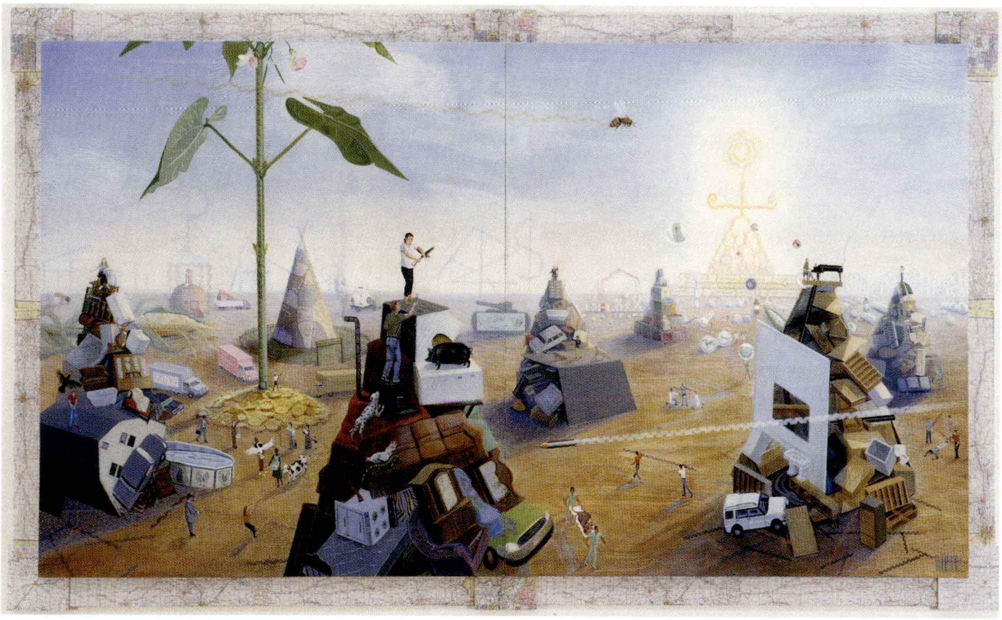

1 Frank Moore, *Oz*, 1999–2000.

IN FRANK MOORE'S *Oz* (1999–2000) (Fig. 1), a colossal genetically engineered plant is embedded in a mound of golden coins, while its environment is littered with piles of defunct furniture, cars, appliances, a farm house, and other sundry products of consumer culture. A huge ear of corn lies in a field, and a large pig stands on the clutter. In this apocalyptic end of the-world scenario, a giant bee leaves a trail of double helices in the sky. On the ground, a prone figure is being beaten by two men with whips configured as strands of DNA. Taking a cue from Hieronymous Bosch's morality tales, Moore, in his meticulously painted visionary style, projects a moral message. He questions the hubris of geneticists and bioengineers. The pot of gold is a lure, but the litter suggests the disasters that can result from such profitable experiments.[1]

The Commodification of Nature and the Body

Genetic engineering techniques transform living things—plants, animals, and human bodies—into marketable biomaterials. Animal bodies are engineered and cloned to serve as research tools. Human body tissue is turned into valuable raw material for pharmaceutical products. Snippets of infant foreskin are used to create artificial skin for burn victims. Drug companies import fetal kidney tissue from Colombia to manufacture Abbokinase, a drug to prevent blood clotting. Human blood in the age of genetics has become a source of clinical and research information. It is today one of the most valuable commodities in the world. Whereas refined petroleum sells for as much as $40 a barrel, an equivalent quantity of blood products is worth about $67,000.

Body tissue, including DNA, has commercial value well beyond medicine. DNA can be used for computing since the four chemicals represented by the letters CATG provide more memory than the binary code of ones and zeros. A British biotechnology firm harvests human sweat to extract pheromones for its product called Xcite—moist towelettes saturated with the sexual hormone that "boost the wearers' sexual smell signature." Nobel Laureate Kary Mullis's company, Stargene, uses PCR amplification technology to create and market jewelry containing DNA cloned from famous rock stars.[2]

References to genetically engineered products in both the scientific and popular literature employ a language of commerce—of banking, investment, insurance, compensation, and patent rights. Pathologists regard banks of stored human tissue samples as "treasure troves" or "national resources" for research. Living things are biosystems that can be divided and dissected down to the molecular level.

Geneticists talk of "prospecting" for genes. Cells, in the language of science, are extracted like a mineral, procured like a parcel of land, harvested like a crop, mined like a resource, patented like an invention, or banked like money. Describing the collection of population-based gene databases in Iceland, its project director, Kari Stefansson, compares them to a cash-dispensing system. From this perspective, the world is a network of centralized databases of personal information tracking finance. Similarly, he argues, gene banks are centralized repositories of personal information on health, and like a cash dispenser, an ATM, they enable the withdrawal of useful and valuable currency.[3] Such language capitalizes the bodies of humans and other animals and transforms them into fungible commodities.[4] "Who owns life?" has become and remains a contentious question facing society in this age of biotechnology as body tissues and fluids, cells, and blood are increasingly commodities—objects of exchange.[5] We remain uneasy, if not reticent, about trading flesh around the world, queasy about keeping corpses alive by artificial means to harvest their body parts, and concerned about genetically modifying animals to use their organs for human transplantation.

The Body Exchange

The economic value of the body, especially as a source of labor, has been a frequent and recurring theme in the visual arts. The production value of the body was the subject of many 19th-century portraits of workers—from the slave to the haymaker to the prostitute to the wet-nurse. Artists such as Courbet, Degas, and Manet explored the ways in which the body's uses were inextrincably tied to social class. Degas, for example, portrays how aristocrats in the 19th century employed the bodies of their servants. *At the Races* (1869) depicts a horse-drawn carriage that halts as a hired wet-nurse breastfeeds the baby. The mother and driver look on. The nurse's milk, her bodily fluids, are a commodity, exchanged for gold. In Manet's *Olympia* (1863), a female courtesan entices her patrons. The bounty surrounding her indicates and measures her commodity value.

The potential for commercial gain from biotechnology applications followed from a series of legislative measures enacted during the 1980s designed to encourage corporate investment in effective research. This legislation allowed universities and nonprofit institutions to apply for patents on federally funded projects while providing tax incentives to corporations investing in academic research. At the same time, changes in patent law turned commercial attention toward research in genetics, blurring the borders between academic and commercial laboratories.[6] Genetics became big business, intrinsically tied to the production of agricultural and pharmaceutical products. And science, beholden to corporate sources of funding and largely driven by the possibilities of patents, has significantly changed.

U.S. patent practices, allowing, for example, the patenting of gene sequences and genetically modified organisms, have been criticized as an affront to human dignity and an abuse of human individuals and animal subjects.[7] How is it possible that living materials can be patented or even copyrighted? Can gene sequences be defined as products or, in the rather loose language of U.S. patent law, "compositions of matter?"[8]

Since the Plant Patent Act of 1930 and the Plant Variety Protection Act of 1970 allow the patenting of life forms, legal scholars have struggled over the relevance of this language to "living things." However, in 1987, the U.S. Patent and Trademark Office issued a policy statement indicating that it considered non-naturally occurring nonhuman living organisms to be patentable subject matter. In applying for gene patents under these provisions, researchers rely on precedents suggesting that naturally occurring substances (such as vitamins) can be patented if they are isolated or altered so as to be applicated in novel ways. Genes, taken from their source, are thus isolated and, therefore, they are not in precisely their natural form. They are, in effect, recontextualized, turned into an artifact, regarded as man-made and, therefore, patentable.

For artists, recontextualization is a familiar process. Just as geneticists turn genes into artifacts, so artists rely on techniques of recontextualization to redefine ordinary objects as art.[9] Take, for example, Duchamp's famous *Fountain* (1917).[10] When placed in the context of a museum, this urinal becomes, not utilitarian hardware, but avant-garde sculpture. Similarly, when Roy Lichtenstein or Andy Warhol transferred comics or advertisements to canvas, these images were viewed as high and serious painting. The meaning of these entities become dependent on activating context, the social and cultural environment. Yet in the precincts of art as well as science, recontextualization is often regarded as an affront. Many people consider works of art as unique, even sacred.

Contemporary Art and the Molecular Enterprise

A vital conjunction of visual artists today question assumptions underlying body commerce. Through their visual images, they are addressing commercialization of the body in commodity culture, the employment of animals as instruments for commerce and/or research, and patenting as well as ownership of genetic material.

The Body in the Commodity Culture

Molecular biologists, writes anthropologist Emily Martin, conceive of the cell as "an assembly line factory in which the DNA blueprints are interpreted and raw materials fabricated to produce the protein end product."[11] The factory is also a salient metaphor for artists. Andy Warhol, for example, named his own studio the "Factory." "This is where I make or *build* my work. In my art work, hand painting would take much too long and anyway, that's not the age we live in. Mechanical means are *today* and using them I can get more art to more people."[12]

Just as the distinction between the laboratory and the factory is dissolving with a growing liaison between business and science, so Warhol's art studio merged with the business of manufacturing. Indeed, capturing the spirit of materialist culture of the 1950s and 1960s, Warhol explicitly identified his works as commodities, equating the success of an artist with commensurate monetary rewards. "Business art," wrote Warhol, "is the step that comes after Art.... After I did the thing called 'art' or whatever it's called, I went into business art. I wanted to be an Art Businessman or a Business Artist."[13]

Warhol expressed his avaricious reverence for cash in an early work, *Untitled (Tattooed Lady)* (1955) (Fig. 2), in effect, a calling card designed for clients of advertising and graphic design agencies.[14] Logos of corporate identity are slavered all over a female figure. Her body is converted into maps of territories, labeled with the pictorial legends of Hunts, Armstrong, Chanel Number 5, and Mobil Oil. In another series of paintings, Warhol depicts American currency in mass-produced serial patterns as an indication of society's infatuation with money.

Producing bodies is the central business of Gunther von Hagens, a physician, anatomy lecturer, and (arguably) an artist at the University of

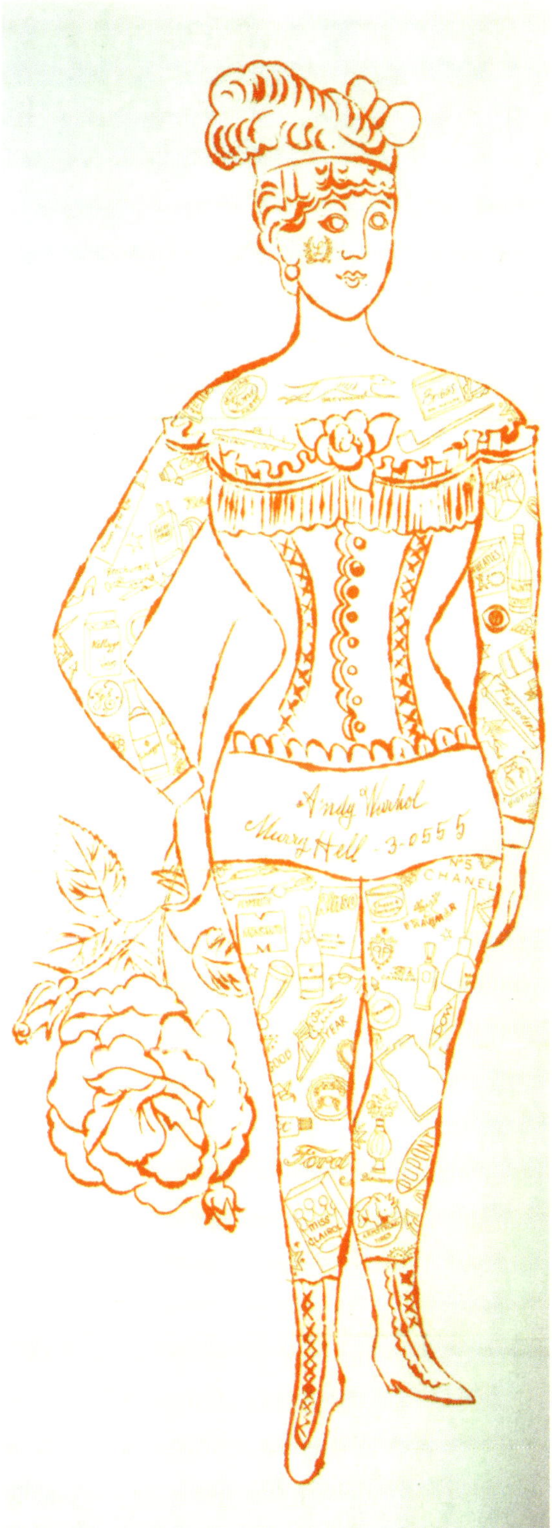

2 Andy Warhol, *Untitled (Tattooed Lady)*, 1955.

Heidelberg. He built a body factory, the Institute for Plastination, devoted to the production of plastinated cadavers. von Hagen's bodies—real human corpses, not sculptures or casts—are prepared through an embalming process, plastination, that preserves body tissue and organs so that they can be displayed in a sort of life-like manner. His bodies, donated to the project by people who, claims von Hagens, saw this as a manifestation of immortality, were originally concocted as visual tools for medical students. They have since been exhibited in both science and art museums around the world in "Body World" shows that are spectacles in the tradition of 19th-century side shows and carnivals.[15] Body figures in these shows are dramatically posed in common positions (running, bicycling, gesturing, praying, playing chess) and colorfully pigmented to reveal blood vessels, muscles, bones, and organs (Figs. 3 and 4). They have no facial expression, and thus appear as cold, emotionless objects; yet they dazzle viewers with their audacity and morbid detail.

von Hagens calls himself an "anatomy artist" while simultaneously insisting that his work illustrates a symbiosis between anatomy and art. But then, he says, whether or not it is art is a matter of interpretation. "I don't see myself as an artist," he breathlessly claims, but he also describes his work as belonging to "the tradition of the artists and scientists of the Renaissance, like Leonardo, Vesalius and Durer [who] embodied the last times when art and science were indistinct."[16] Many of his body displays appropriate their themes from historical art sources—a portrayal of a body with doors and drawers resembles Dali's Drawer Man, a horse and rider echoes the 18th-century work of anatomist and artist Honore Fragonard, and a man with his flayed skin over his arm resembles many classic 16th- and 17th-century figures: Bartholomew's painting in Michelangelo's Sistine Chapel, Giulio Casserio's late 16th-century engraving of a figure peeling back his skin to reveal his viscera, and John Browne's 1681 sketch of a body flaunting his exposed muscles and flayed skin.[17]

von Hagens seeks and receives a significant amount of media attention. He created a TV spectacle in England by conducting a public autopsy in the gallery where "Body World" was exhibited. Under a copy of Rembrandt's "Anatomy Lesson of Dr Nicolaes Tulp" (1632), before an audience that paid admission fees, he dissected a cadaver on a table. The drama attracted the anticipated press. von Hagens prominently posts inflammatory news articles concerning his work at the entrance to his shows. Headlines raise questions: "Would you Call this Art?" "Artist or Butcher?" "It's Dead but is it Art?" von Hagens responded: "I use the word 'art' very cautiously because it has come to mean business and entertainment." But his own enterprise is nothing if not commercial. It is entertaining and even sensational in the tradition of the early freak shows. He has patented his process of plastination, and sells souvenirs, stationery, and even T-shirts at his exhibitions.

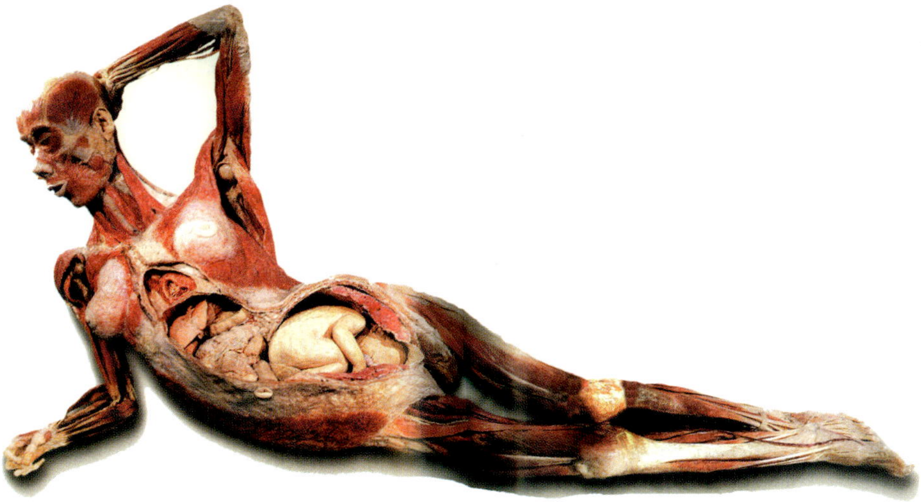

3 Gunther von Hagens, Reclining Woman in the 8th Month of Pregnancy.

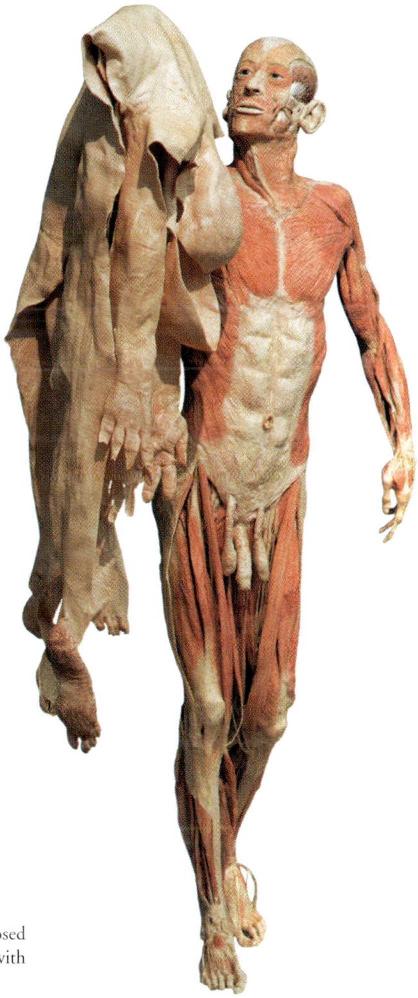

4 Gunther von Hagens, Posed Whole Body Specimen with Skin.

If plastinated bodies represent the commercialization of the corpse, Marc Quinn's *Self* (1991) represents the commodification of blood (Fig. 5). *Self* is a bust of his head composed entirely of 9 liters of the artist's congealed, frozen blood. This sculpture is kept refrigerated in a glass cabinet and is part of the Saatchi collection. Every five years, Quinn creates a new version using fresh blood. Quinn, describing the motive for this work, says he is critical of "the over-determined factuality of biological research.... Do we only think we know what the natural is because of what we have found through science?"[18] By placing his own biomaterials on display in an art gallery, Quinn also addresses how body materials have been commodified in both science and art as blood and tissue become both commercial biotechnology products and a medium for artists as they pursue new forms advancing the context of the spectacular.[19]

For example, conceptual artist Chrissy Conant displays and then sells her eggs in an exhibit entitled *ChrissyCaviar*® (2001–2002). Undergoing hormone fertility treatment so as to "super-ovulate," producing multiple eggs, she harvested and packaged twelve of them as if they were hens' eggs commercially packaged by the dozen. These eggs were then placed in sealed vials resembling Beluga caviar jars, labeled "Caucasian: Packed by Private IVF Center, USA. Product of Conant Ovaries, Keep Refrigerated." A seductive image of the artist reclines on the jar lid (Fig. 6), and each jar is signed and up for sale. Conant expects that her product will "surpass Beluga caviar as the current ultimate in luxury consumable items. In the context of fine art, using my genes as a commodity, I am making art with my body." Also describing her work as a marketing ploy to capture a male: "I have no objection if the purchaser wants to pop open the jar and eat it," says Chrissie.[20]

In 1989, British artist Rick Gibson created his own spectacle—a sculpture called *Human Earrings*, intended as a comment on how we treat fellow human beings.[21] Gibson obtained two fetuses of about 3–4 weeks' gestation from a professor of anatomy. He then sculpted a head and hung the freeze-dried fetuses as earrings. He was prosecuted for the common law offense of outraging public decency. Both the artist and the gallerist owner were fined.

Commodification of women's bodies has consistently been both a genre and motif in art, but in the genetic age, increasing possibilities of manipulation have added novel dimensions. The Barbie doll is a striking commercial icon representing an ideal and now manipulable woman's body. Barbie was actually based on a German Lilli doll that was in turn modeled after a pornographic caricature in a German comic strip. She became the product of global assembly lines, designed in Los Angeles, manufactured in Malaysia, and distributed throughout the world. Barbie quickly ascended to a metaphor in popular culture and art for a commodity culture in which persons—especially young women—are objects to be manipulated, engineered, and marketed. Barbie is also an unintended parody on "technically-assisted modification" through cosmetic surgery, and now through genetic modification as

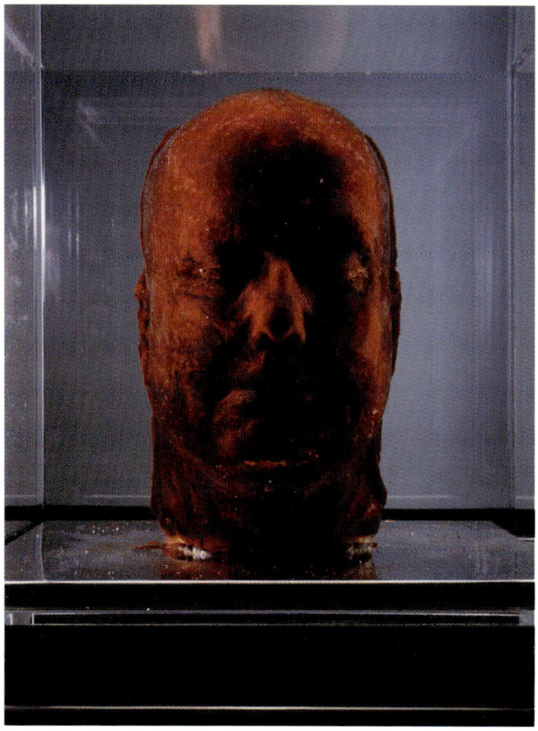

5 Mark Quinn, *Self,* 1991.

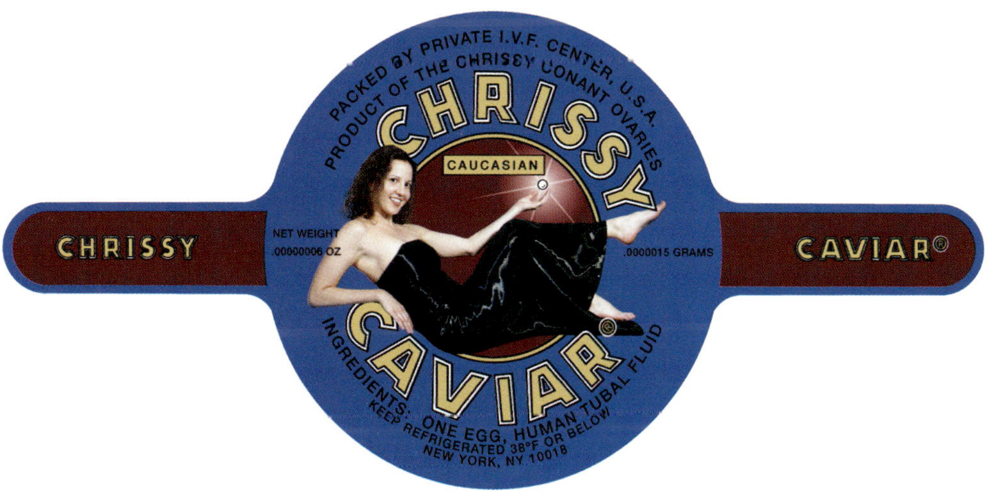

6 Chrissy Conant, *Chrissy Caviar*®, 2001–2002.

the body becomes "raw material to be fragmented into parts, molded, and reshaped into perfect form."[22]

Biological iconography has become a suitable visual image in another commercial sector, the design world of haute couture. Helen Storey is a fashion designer whose clothing takes its visualization from biological processes. Collaborating with her sister, Dr. Kate Storey, a research biologist at Oxford University, she created a collection of 27 garments entitled *Primitive Streak*, recalling the first 1000 hours of embryonic development following fertilization (Figs. 7 and 8).[23] Her garments, exhibited not as fashion but as high art, sport designs of neurons, DNA sequences, limb buds, and even heart tissue. Just as repeated patterns of tractors and factories appeared on the textiles of the early Russian avant-garde, so biological images in Storey's fabrics reflect the ideology and the visions of today's techno-driven desires.

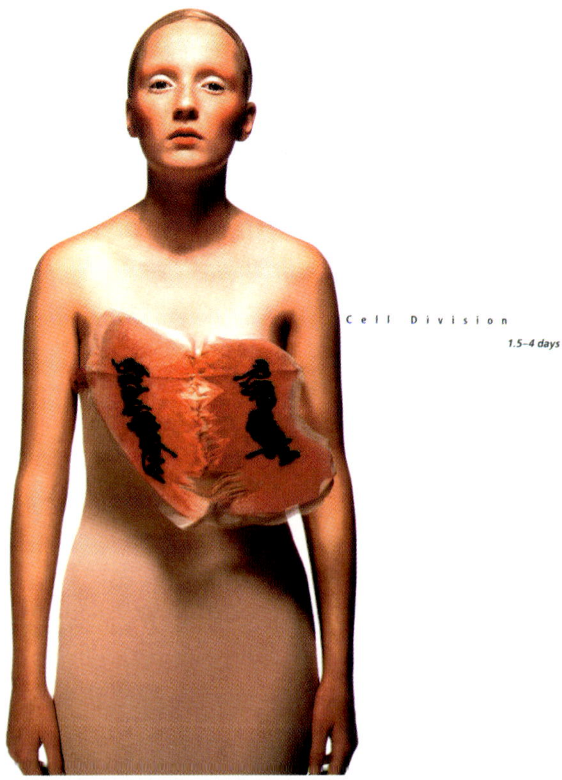

7 Storey Sisters, *Design 11: Anaphase Dress* (from *Primitive Streak),* 1997.

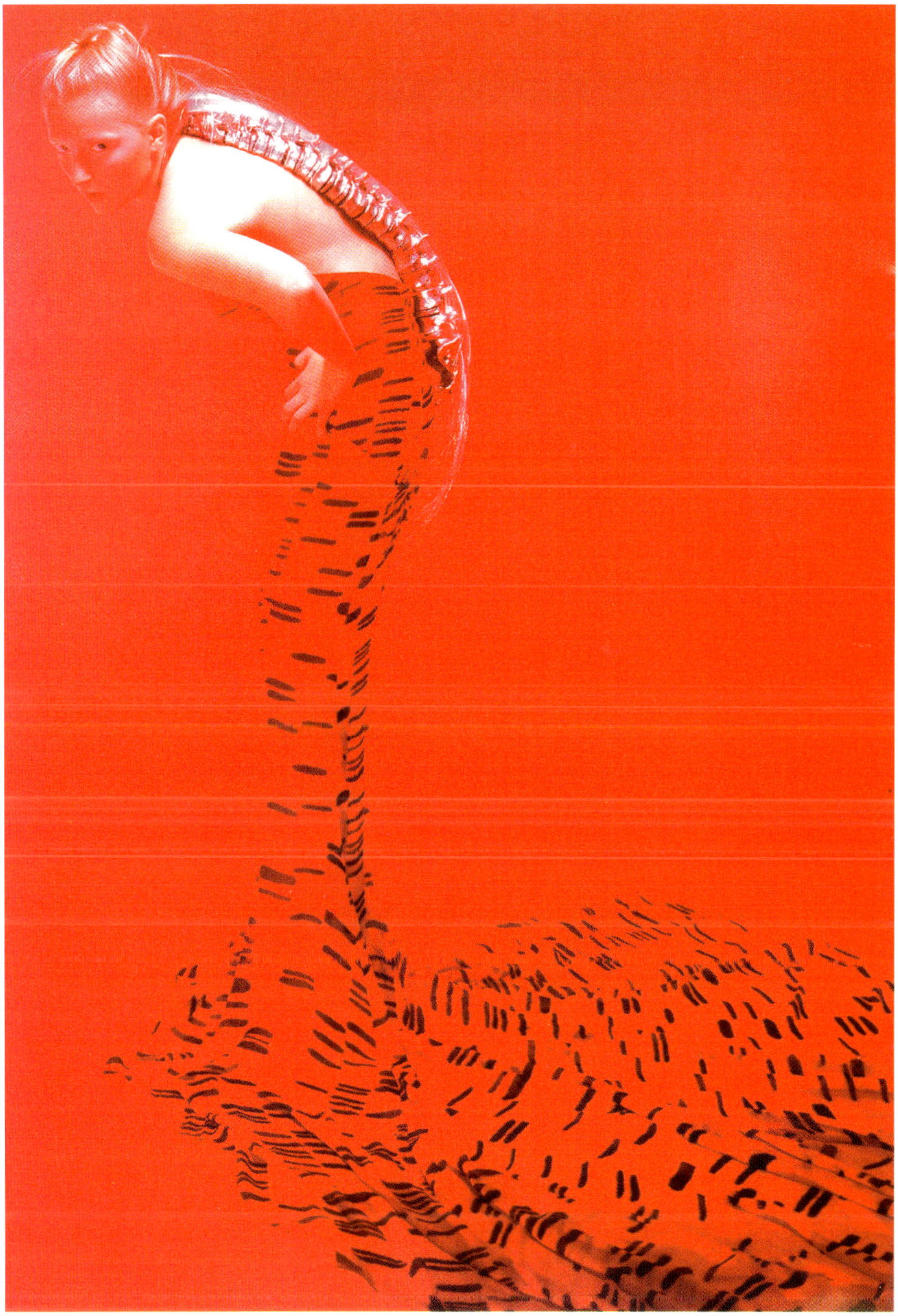

8 Storey Sisters, *Design 26: DNA Sequence/Spinal Column Dress in Red Silk Chiffon* from *Primitive Streak,* 1997.

The Animal as Instrument

Animal models are routinely used in scientific laboratories for the study of genetic diseases. Laboratory mice have been among the most significant models for human disease research since Clarence Cook Little developed the first inbred mouse strain in 1909. Today, gene-altering stem cells are introduced into mouse embryos such that the mice will carry a specific defect. Such mice are essential tools that assist researchers to better understand the symptoms of disease and to discover therapeutic solutions. In April 1988, the U.S. Patents and Trademarks Office (PTO) granted a patent for a mammal, a mouse that had been genetically engineered with human genes to serve as a model for cancer research. This oncomouse, with an inserted human gene sequence conferring susceptibility to cancer, was the first patented genetically engineered animal in the world. U.S. Patent No. 4,736,866 was issued to two researchers from Harvard University. Harvard then licensed the patent to DuPont, the company sponsoring the research.[24] The Charles River Laboratory in Massachusetts marketed these mice at costs ranging from $50 to $75.[25] Living instruments for research, these confected creatures populate numerous research laboratories.

The creation of mice as animal models for research attracted the early interest of Frank Moore. In *Wizard* (1994) (Fig. 9), four oversized laboratory mice, walking on a field of pills, follow a scientist who carries laboratory notebooks. The scientist, recalling the image of a Pied Piper, gazes over his shoulder at his rodent entourage.

Oncomouse is also the subject of Bryan Crockett's huge *Ecce Homo* (2000) (Fig. 10). It stands six feet tall, is hairless as is the living oncomouse, and is intended, Crockett writes, as an interpretation of "the ultimate figure of salvation, Christ, through the ultimate actor of contemporary science, the Oncomouse." His figure is "a monument to the test object of modern science, human-kind's symbolic and literal stand-in personified."[26] Science, for Crockett, is displacing religion.[27]

In a 2002 show entitled "Cultured,"[28] Crockett exhibited seven smooth-skinned baby pink mice cast in marble dust and resin. His mice are caricatures, appearing as personifications of the seven deadly sins: pride, envy, gluttony, anger, lust, greed, and sloth (Fig. 11). Each is a genetically engineered creature. *Envy* has human ears, referring to ears that have been grown experimentally on mice. *Gluttony* has been given a gene for obesity, representing the genetically engineered mouse used to study diabetes. *Lust* has been engineered with a delicately sensitive skin. Crockett's humorous medley of rodents (recalling, perhaps, the seven dwarfs) is a critique of the creation and patenting of living creatures, and an ironical comment on the instrumental use of animals to study human pathologies.

Other "recipes" for rodents are offered up by Catherine Chalmers's photographs. Commissioned by *The New York Times* magazine in 2000,

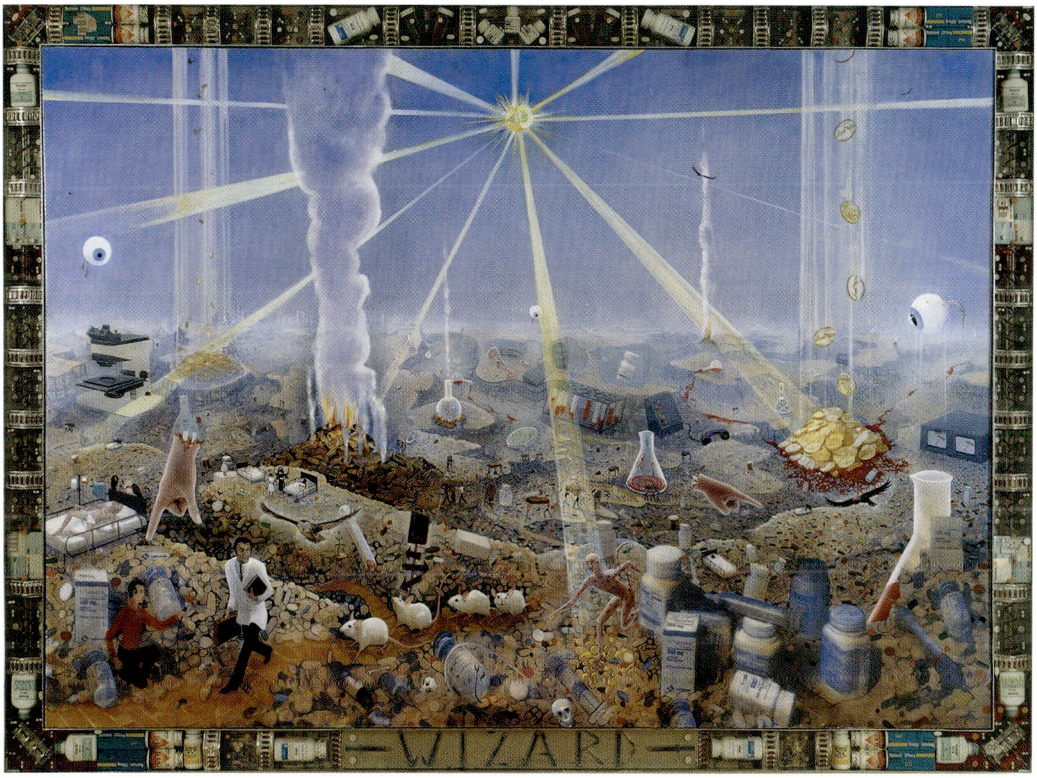

9 Frank Moore, *Wizard*, 1994.

Chalmers photographed a number of mouse models (Fig. 12) at the Jackson Laboratory in Bar Harbor, Maine, for an article wittily entitled, "Fuzzy Little Test Tubes."[29]

The utility of the earliest animal patents had mostly to do with their experimental value. Subsequently, the PTO received patent applications for animals genetically engineered to serve as living "factories" for the production of pharmaceuticals and other commercial products.[30] Goats, for example, have been engineered with human genes to express monoclonal antibodies or several other large protein drugs in their milk. Two thousand such goats reside at a "biopharm production facility" in Massachusetts.[31]

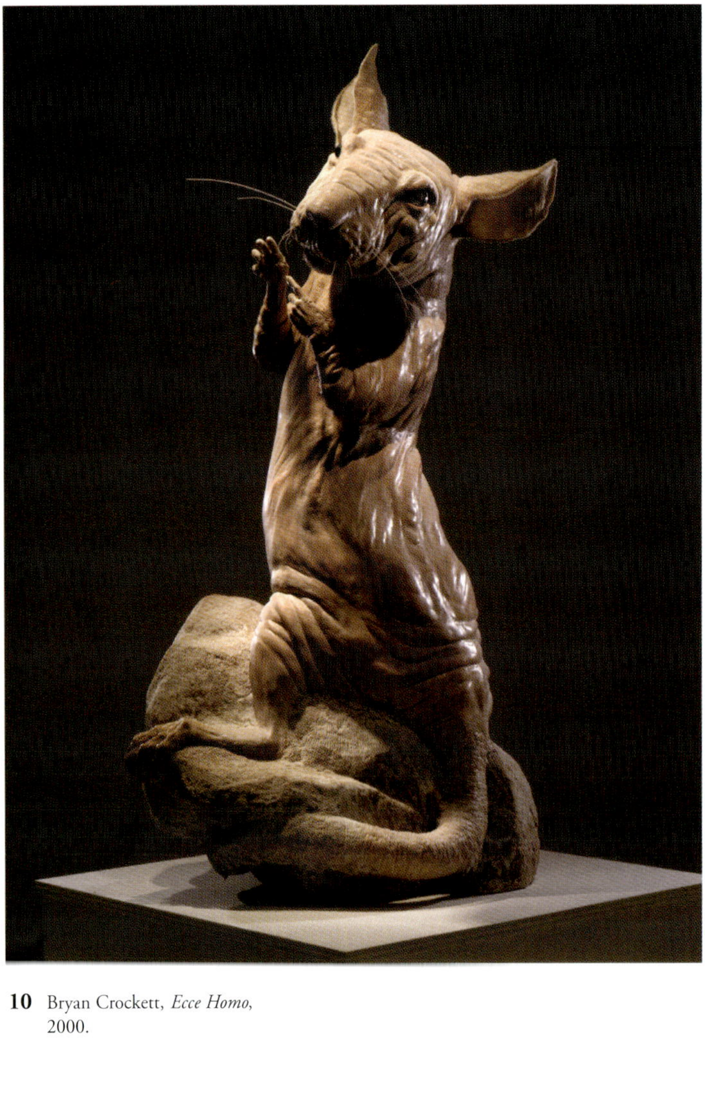

10 Bryan Crockett, *Ecce Homo,* 2000.

In 2002, an American subsidiary of PPL Therapeutics, the Scottish firm that cloned Dolly the sheep, followed up its earlier transgenic experiments (see earlier chapter, Mutation, Manipulation, and Monsters) by successfully cloning four piglets with organs genetically altered for human transplantation. Genes that would trigger rejection of transplanted organs had been "knocked out," opening the possibility of commercially producing pigs that would have transplantable organs. This process, intended to resolve the shortage of human organs, was regarded as commercially viable. Significantly, the company revealed its experiment to the public, not in a peer-reviewed scientific journal, but in a stock market offering as an advance in "the regenerative medicine business."[32]

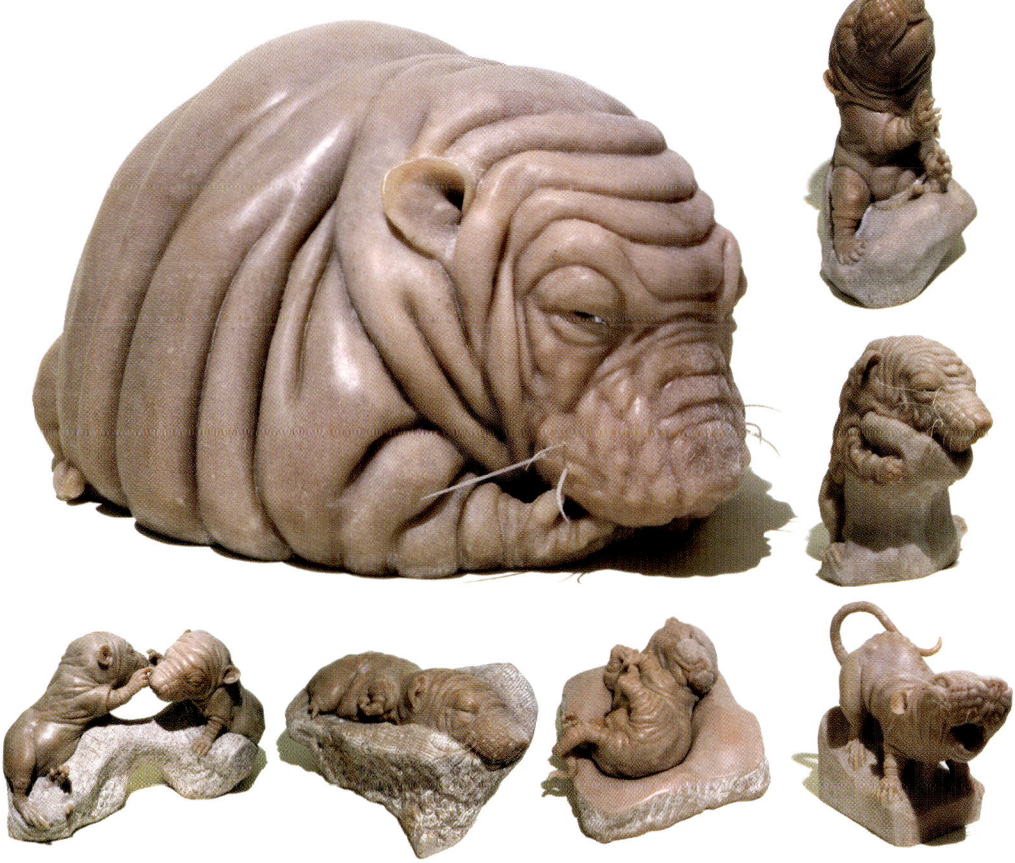

11 Bryan Crockett, *Pride, Lust, Gluttony, Sloth, Greed, Envy, Anger,* 2001.

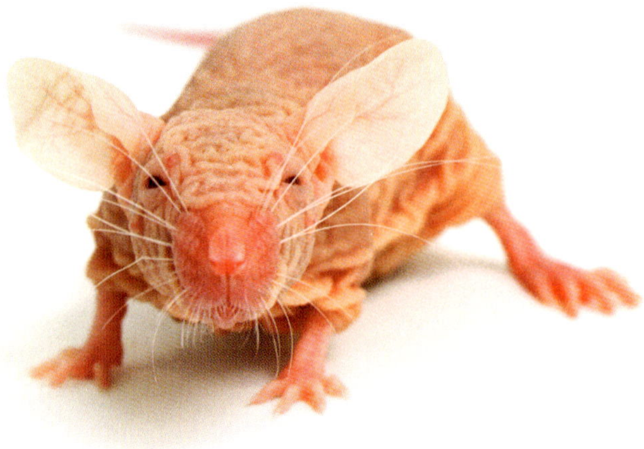

12 Catherine Chalmers, *Rhino*, 2000.

Frank Moore, in *Cocktail* (2001) (Fig. 13), explicitly addresses the commercial aspects of the genetic modification of both animals and plants. A syringe containing a fish and a chicken injects a genetic cocktail into corn—perhaps the most controversial of all genetically modified crops. The plunger mimics the structure of the double helix, two coins, and a pack of dollar bills. "Cocktail" graphically and dramatically depicts the role and driving force of money in developing techniques of genetic modification.

Debates over the marketing of genetically modified foods are also the focus of Christy Rupp's sculptures and installations. She addresses the controversial labeling requirements in *New Labels for Genetically Engineered Food* (2000) (Fig. 14). This installation consists of plastic food containers arranged in a spiral with labels such as "Original DNA Product" and "Randomly Mutating Food." She notes ironically that revealing labels are conspicuously absent in the genetically modified products sold in supermarkets in the United States, while they are required in many European countries.

Genetically engineered food is also the subject for David Bowers' double helix icon composed of animals and vegetables—chicken legs, lobsters, peppers, leeks, peas, and tomatoes.[33] An incongruous figure, it recalls the 1920s surrealist collages of Max Ernst and Hannah Höch (see earlier chapter, Mutation, Manipulation, and Monsters), and like these collages, it is a social commentary on the violation of natural order.

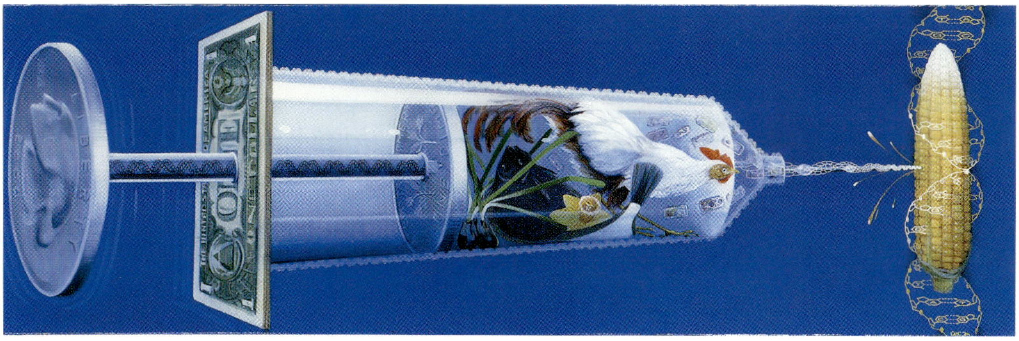

13 Frank Moore, *Cocktail*, 2001.

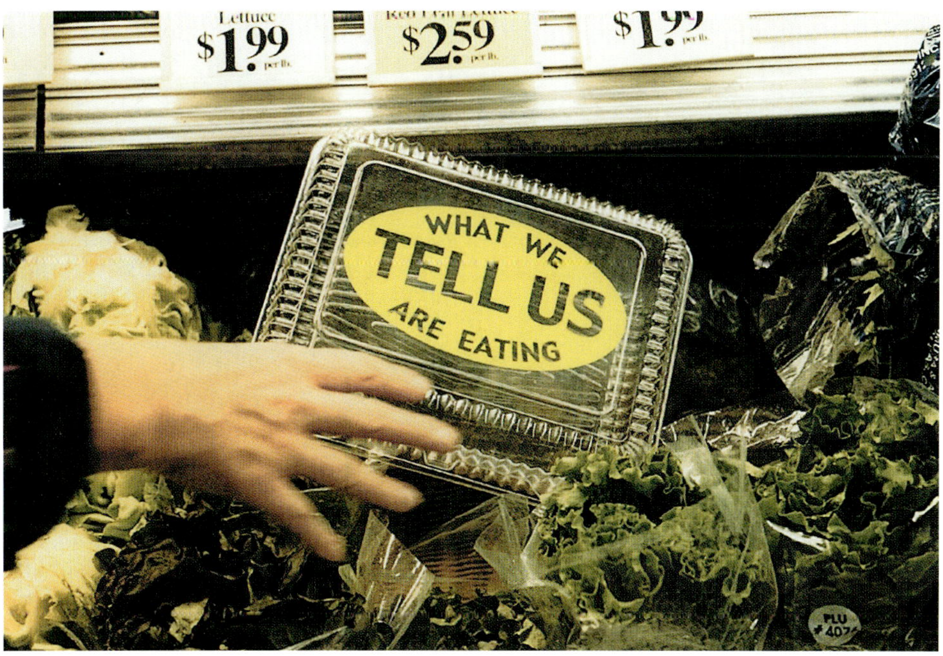

14 Christy Rupp, *New Labels for Genetically Engineered Food*, 2000.

Just as genetically modified crops have been developed to withstand varied climatic conditions, so too have some animals. For example, in 2001, research scientists at Hebrew University, Israel, bred a featherless chicken designed to remain cool in hot climates, hence reducing the need for water and electricity in processing them for consumption (Fig. 15).[34]

The genetic engineering of animals for commercial purposes was the central focus of Laurel Katz's 1994 installation, *The Spotted Merino/Hooked Beetle Sweater Production System* (1994) (Fig. 16). Katz invented a "hooked beetle" genetically engineered to have legs resembling hooks. She imagines this beetle trained to walk in circles on the backs of sheep catching their wool in its leg hooks and looping the yarn to weave sweaters for the market. In this parasitic relationship, animal integrity gives way to industrial design. Her installation critiques the genetic manipulation of natural processes for commercial ends.

Katz's far-out fantasy, however, is, in fact, not that distant from our current reality. Laboratory scientists and fellow-traveling entrepreneurs have been scheming for years to discover ways to transfer the silk-making genes of spiders into organisms that will produce silk for mass production. A Canadian company, Nexia Biotechnologies, has reported their success in

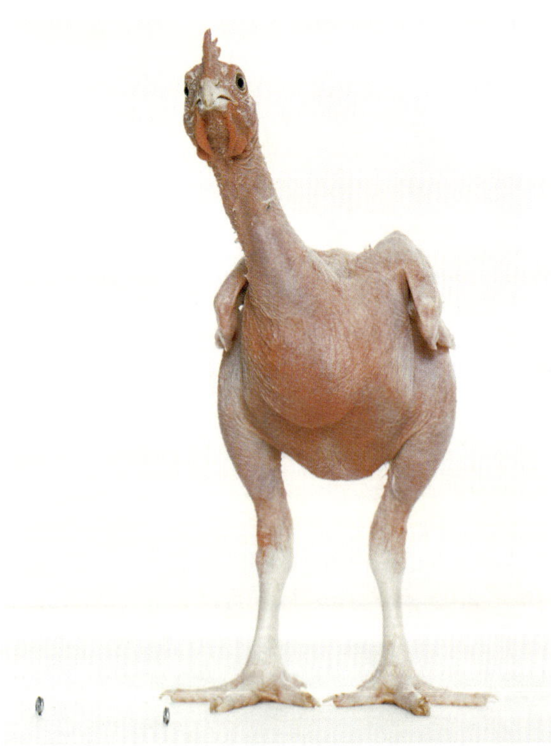

15 Adi Nes, *Untitled (The Featherless Chicken)*, 2002.

splicing silk genes into mammalian cells that mimic those integral to the spider itself. They then splice the silk genes into bovine mammary cells and hamster kidney cells. Both cell lines now secrete soluble silk proteins. The geneticists then harvest recombinant proteins from these cells, spinning them into long, lightweight and silk fibers now being marked as BioSteel.[35] Conversely, Japanese researchers have modified silkworms to produce a mammalian protein—a human-type collagen—in their cocoons. Collagen is highly useful in tissue engineering as well as drug delivery systems and other medical applications.

Who Owns Whose Body?

Ever since a 1980 U.S. Supreme Court Decision (Diamond v. Chakrabarty) allowed the patenting of a live, genetically engineered microorganism, researchers have been applying for patents on cell lines and genes. The Chakrabarty decision applied only to the microorganism that was defined by Chief Justice Burger as constituting a "manufacture or com-

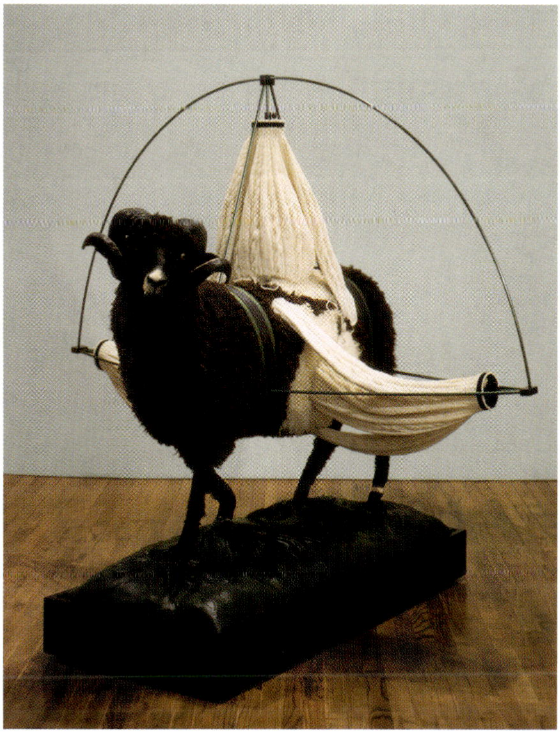

16 Laurel Katz, *The Spotted Merino/Hooked Beetle Sweater Production System*, 1994.

position of matter" and therefore a patentable subject matter. But it also laid the groundwork for future patent practices that are gradually extending and enhancing the definition.

Under U.S. patent law, DNA and other human tissue—once isolated and altered—can be bought, sold, licensed, and owned.[36] Recent patent applications include projects that could lead to the creation and ownership of persons. In April 2001, for example, the U.S. Patents and Trademarks Office granted patent number 6,211,429 to the University of Missouri for a "method for transplanting a nucleus from a cultured mammalian cell, mammalian embryos, mammalian fetuses, or an adult mammal to a recipient mammalian oocyte; producing a cloned mammalian embryo and producing a cloned mammal."[37] The patent encompasses the living, cloned products that are created by the method. These products would not exclude humans.

The technical possibilities of producing such a complex mammal are daunting and unlikely, but such patent applications, including those for embryos and fetuses, raise troubling questions that are the focus of considerable philosophical and legal debate—enhanced by visual representation.[38] Is it ethically and ecologically appropriate to patent a product of nature, even when it is altered to meet the constraints of patent law? Are presuming scientists directing the course of evolution? What are the implications for human dignity, for the value of life itself, if the body becomes a commodity?

A dramatic landmark case in 1990, *Moore v. University of California,* brought these questions to public attention. John Moore, a Seattle businessman, fell ill with a rare form of hairy cell leukemia and had his spleen removed at UCLA's Medical School. After 7 years of repeated tests and biopsies, Moore discovered that his spleen cell line (the Mo-line) had become patent number 4,438,032. He sued his doctors for malpractice and property theft. Considering Moore's case, the California Supreme Court denied Moore's claim that he owned his own body tissue and allowed the UCLA scientist and his biotechnology firm to obtain patent rights, arguing that the cell line is "factually and legally distinct from the cells taken from Moore's body" and that it was essential that "we not threaten... innocent parties who engage in socially useful activities such as researchers." If the sources of human tissue were to assert their rights, said the court, this could be detrimental to the "infant biotechnology industry as companies are unlikely to invest heavily in developing, manufacturing, or marketing a product when uncertainty about clear title exists."[39] The future or failure of biotechnology was at stake. The idea expressed in the court's decision—that a scientist or company can patent parts of the human body—was counterintuitive and shocking to many observers. The Mo-line, as the cells were called, seemed to represent the ultimate commodifying of the body. If it is deemed acceptable to patent human genes and sell them, why not market human organs, or even babies, or persons?

The idea that a person can be treated as a commercial object inspired Larry Miller to address the issue in his 1992 elegant, formal, and official-looking certificate: *Genetic Code Copyright Certificate* (1992) (Fig. 17a). The certificate states that "I, a naturally born human being do hereby forever copyright my unique genetic code, however it may be scientifically determined, described or otherwise empirically expressed." Miller's work plays on the construction of DNA as the essence of an individual human being, the underlying source of identity. He is mocking the idea that an individual's genetic code can be copyrighted and commodified as if a person is no more than a widget. His sardonic comments even go further. In *Genomic License* (Fig. 17b), he refines his business acumen to licensing agreements. No commercial possibility will be left undone.

Yet, here too, real life has kept abreast of art. A British woman, angered at the practice of patenting human genes, applied for a patent on herself, claiming that she is both novel and useful. "It has taken 30 years of hard labor for me to discover and invent myself and now I wish to protect my invention from unauthorized exploitation."[40]

Anticipating a probable trend, an American company, the DNA Copyright Institute Inc., set up a business to copyright DNA because in the future, according to the company's founder, it will be impossible for celebrities to avoid being cloned and copied. For $1500 the company will take DNA samples, store them, and have a genetic fingerprint ready just in case legal action is called for. "Our purpose is to insure that all citizens and their DNA profiles are generally covered under existing copyright laws."[41]

Ellen K. Levy derived a series of digital prints from the diagrammatic drawings that are part of actual patent applications. *Recombinant + DNA* (2002) (Fig. 18) is a double scroll, narrow, and vertically oriented displaying data, graphs, and symbolic notations that highlight the patenting of new life forms. In other works, Levy employs texts from patent applications for products ranging from a gene for "increased persistence" to the methods and apparatus for selling "an aging food product." For Levy, biopatents are a natural extension of the ways in which intellectual property is conceived, protected, and catalogued.

Christine Borland's *Hela* (2000) exposes the problems of the ownership of body parts. In 1951, cells had been taken and cultured from the rapidly growing ovarian cancer of an African-American woman, Henrietta Lacks, and named HeLa cells. These cells possess a rare genetic anomaly that causes them to multiply at an astonishing rate and to survive in vitro, that is, outside the body. They are commercially available and enormously profitable because they are used by scientists throughout the world as the basis of cell research. The ownership of these cell lines became controversial when, in 1975, the Lacks family, who had not been aware that the cells were being used and had never consented to their use, discovered their value from a

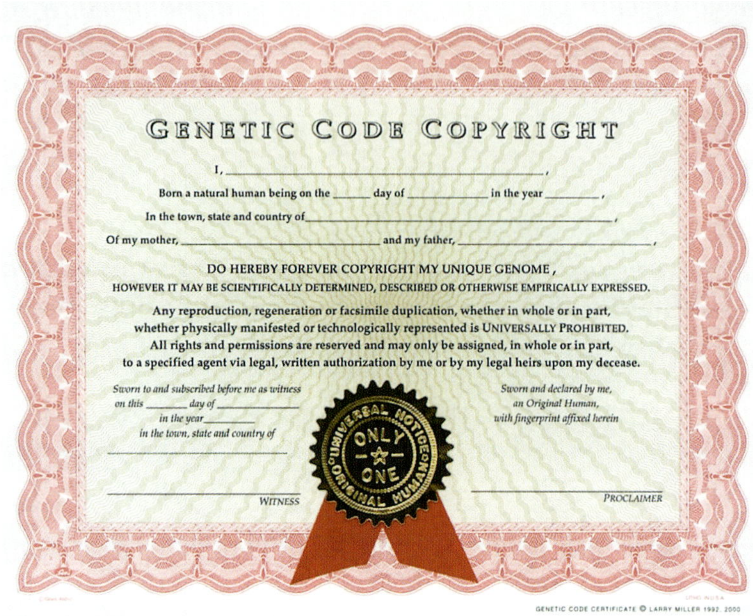

GENETIC CODE COPYRIGHT

I, ____________________,

Born a natural human being on the ______ day of ____________ in the year ________,

In the town, state and country of ____________________,

Of my mother, ____________________ and my father, ____________________,

DO HEREBY FOREVER COPYRIGHT MY UNIQUE GENOME,

HOWEVER IT MAY BE SCIENTIFICALLY DETERMINED, DESCRIBED OR OTHERWISE EMPIRICALLY EXPRESSED.

Any reproduction, regeneration or facsimile duplication, whether in whole or in part, whether physically manifested or technologically represented is UNIVERSALLY PROHIBITED. All rights and permissions are reserved and may only be assigned, in whole or in part, to a specified agent via legal, written authorization by me or by my legal heirs upon my decease.

Sworn to and subscribed before me as witness on this ______ day of ____________ in the year ________ in the town, state and country of

____

____________________ WITNESS

Sworn and declared by me, an Original Human, with fingerprint affixed herein

____________________ PROCLAIMER

GENETIC CODE CERTIFICATE © LARRY MILLER 1992, 2000

17a Larry Miller, *Genetic Code Copyright Certificate,* 1992.

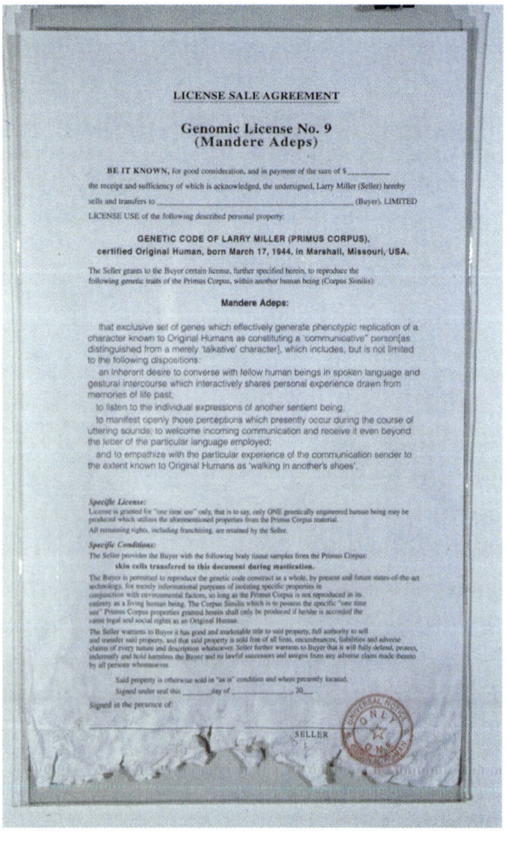

LICENSE SALE AGREEMENT

Genomic License No. 9
(Mandere Adeps)

BE IT KNOWN, for good consideration, and in payment of the sum of $__________ the receipt and sufficiency of which is acknowledged, the undersigned, Larry Miller (Seller) hereby sells and transfers to ____________________ (Buyer), LIMITED LICENSE USE of the following described personal property:

GENETIC CODE OF LARRY MILLER (PRIMUS CORPUS), certified Original Human, born March 17, 1944, in Marshall, Missouri, USA.

The Seller grants to the Buyer certain license, further specified herein, to reproduce the following genetic traits of the Primus Corpus, within another human being (Corpus Similis):

Mandere Adeps:

that exclusive set of genes which effectively generate phenotypic replication of a character known to Original Humans as constituting a 'communicative" person[as distinguished from a merely 'talkative' character], which includes, but is not limited to the following dispositions:

an inherent desire to converse with fellow human beings in spoken language and gestural intercourse which interactively shares personal experience drawn from memories of life past;

to listen to the individual expressions of another sentient being;

to manifest openly those perceptions which presently occur during the course of uttering sounds; to welcome incoming communication and receive it even beyond the letter of the particular language employed;

and to empathize with the particular experience of the communication sender to the extent known to Original Humans as 'walking in another's shoes'.

Specific License:

License is granted for "one time use" only, that is to say, only ONE genetically engineered human being may be produced which utilizes the aforementioned properties from the Primus Corpus material.

All remaining rights, including franchising, are retained by the Seller.

Specific Conditions:

The Seller provides the Buyer with the following body tissue samples from the Primus Corpus:

skin cells transfered to this document during mastication.

The Buyer is permitted to reproduce the genetic code construct as a whole, by present and future states-of-the-art technology, for merely informational purposes of isolating specific properties in conjunction with environmental factors, so long as the Primus Corpus is not reproduced in its entirety as a living human being. The Corpus Similis which is to possess the specific "one time use" Primus Corpus properties granted herein shall only be produced if he/she is accorded the same legal and social rights as an Original Human.

The Seller warrants to Buyer it has good and marketable title to said property, full authority to sell and transfer said property, and that said property is sold free of all liens, encumbrances, liabilities and adverse claims of every nature and description whatsoever. Seller further warrants to Buyer that it will fully defend, protect, indemnify and hold harmless the Buyer and its lawful successors and assigns from any adverse claim made thereto by all persons whomsoever.

Said property is otherwise sold in "as is" condition and where presently located.

Signed under seal this ______ day of ____________, 20___.

Signed in the presence of:

____________________ ____________________ SELLER

17b Larry Miller, *Genomic License.*

friend who just happened to be a scientist. Later the family was asked to provide further DNA samples because the original cells had contaminated other cell lines. But the ownership of cells from an African-American woman inevitably hinted of slavery. Moreover, the family regarded Henrietta Lacks, not as a valuable object, but as a human being—as their intimate kin.[42]

Australian media artist Justine Cooper employes light microscopy images of blood, pus, phlegm, cervical mucus, and tears in her photographs and video installations. In *Moist* (2002) (Fig. 19), she creates microuniverses of untapped colonies, whose sublime countenance opens up future worlds and markets.

The value of cells and genes is reflected in the development of extensive commercial gene data banks, created for epidemiological studies of common disorders so as to develop targeted therapies.[43] DeCode Genetics, an Icelandic biotechnology company created by molecular biologist entrepreneur Kari Stefansson, has collected and banked population-based genetic databases and genealogies from the entire Icelandic population. The company gained the right to store and commercialize the genes, and it sells access to such data to pharmaceutical firms. Similar projects are already under way in Estonia and Tonga, and are proposed for the United States and Great Britain. In effect, entire nations are putting their people's genomes on the block.[44]

Catherine Wagner addresses the banking of human tissue samples in a series of photographs entitled *–86 Degree Freezers: 12 areas of concern and crisis* (1995) (Fig. 20). This work was commissioned by the Tokyo branch of Comme des Garcons for this designer fashion shop's window display. Wagner's photos depict a storage facility, with row upon row of packaged biological specimens and many stacks of trays and bottles. The contents are all labeled as the genetic material collected to study diseases ranging from cancer to HIV, from Alzheimers to bipolar disorder, but it reminds one of a supermarket shelf or a mass-production enterprise.

Wagner's image vividly recalls the huge repositories that store tissue specimens. For example, the repository at the Armed Forces Institute of Pathology in Washington DC, established in 1917, stores millions of tissue specimens that were sent to them for clinical analysis. The American Type Culture Collection lists thousands of people's cell lines that are available for sale; for example, CRL 5867, a 49-year-old black female with cancer of the lymph node, and SK-HEP-1, a 52-year-old German man with liver cancer. Such data, once visible only to scientific and medical professionals, are now represented as art: They have expanded into both the commercial and cultural domains.

Genes and Art for Sale

Exploring the idea of turning tissue and animals (and potentially people) into products, artists have created images that represent the commercializa-

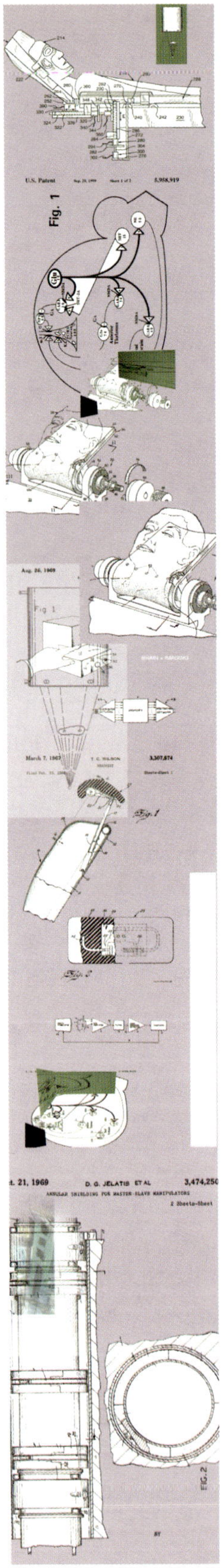

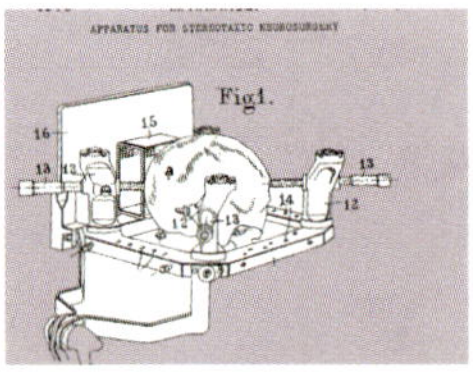

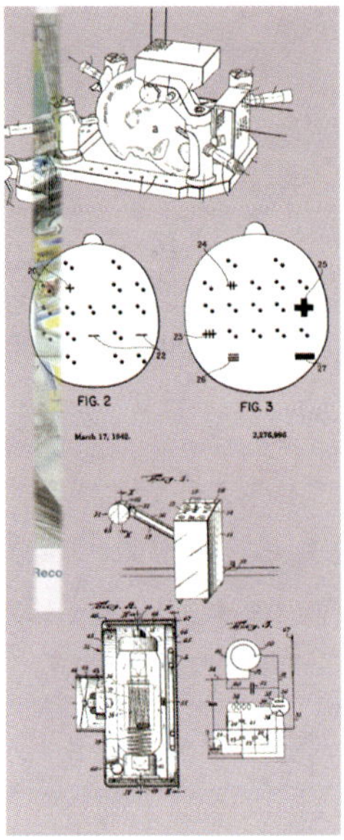

18 Ellen Levy, *Recombinant + DNA*, 2002.

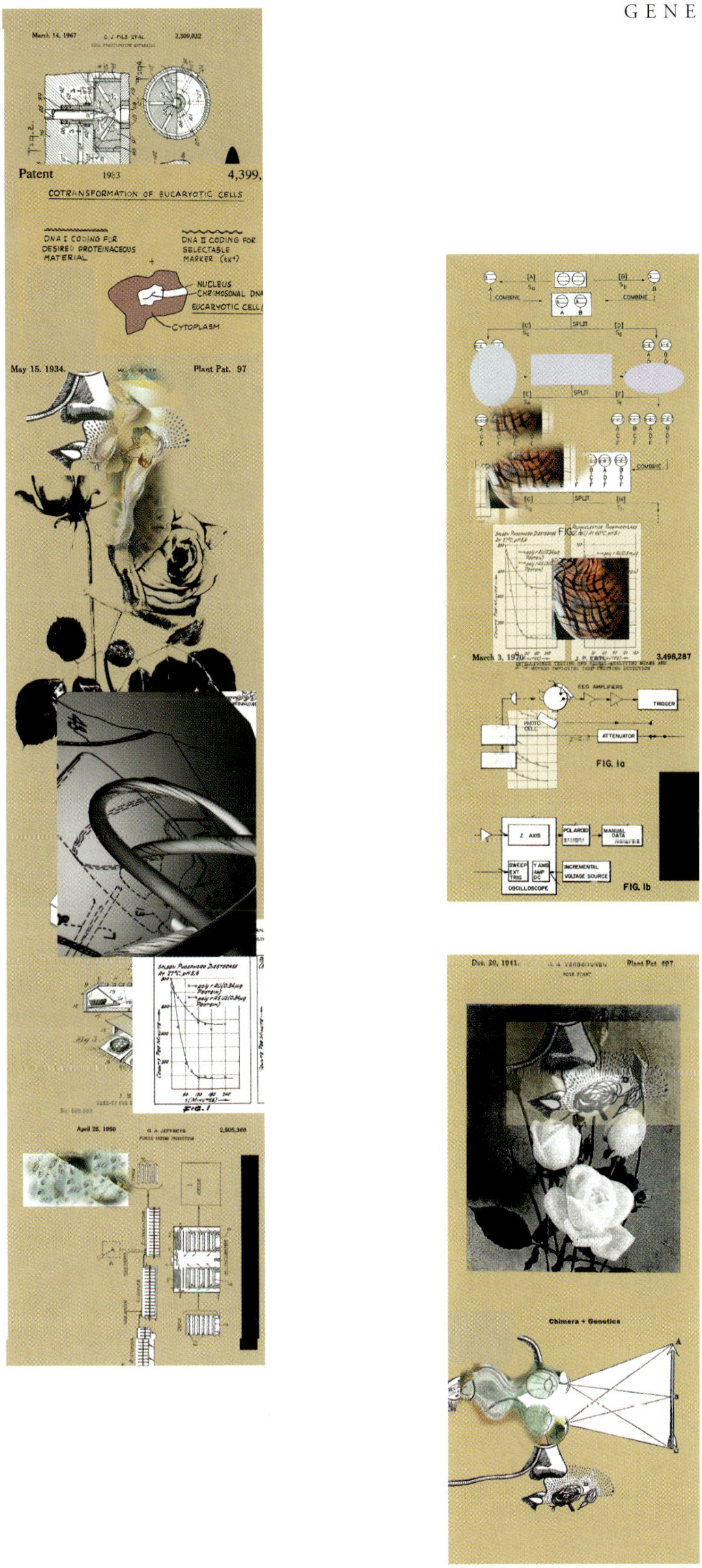

18 Ellen Levy, *Recombinant + DNA*, 2002.

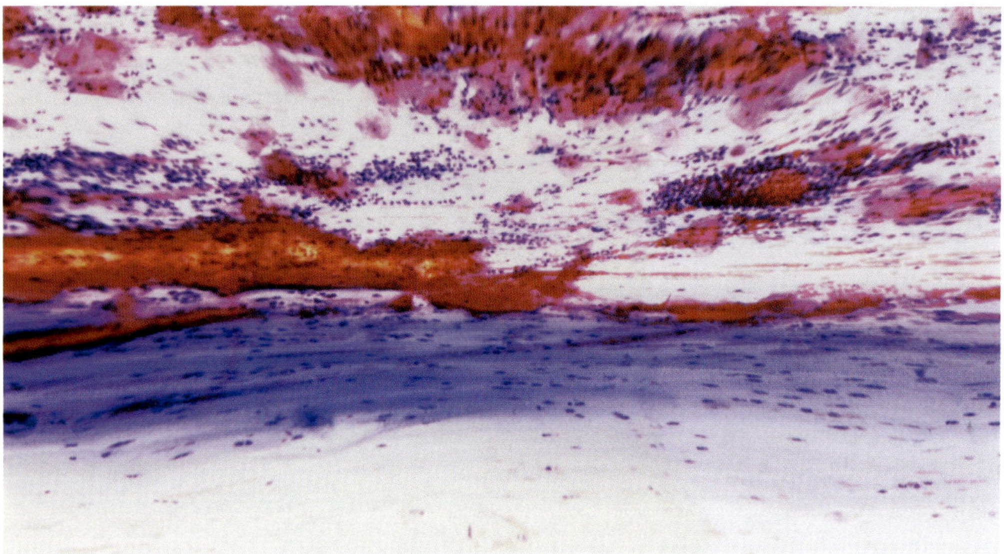

19 Justine Cooper, *Moist* (video still), 2002.

tion of science and the commodification of the body in the biotechnology epoch. This, we first thought, is an unlikely subject for the visual arts, but the concept of the body as a commercial object provides some striking images that convey common assumptions about the value of human life and current anxieties concerning even research and development.

If the artists' spin on transgenics reflects anxieties about authenticity and species integrity, their approach to commercialization reflects concerns about the intrinsic value of life itself when research on the body yields patents and subsequent profits. To treat the body as a commodity like a car, an appliance, or a mousetrap offends a widening spectrum of cultural sensibilities. The components of the body and its tissues are not just neutral matter. The body and its parts are layered with cultural significance, religious ideologies, and personal associations.[45] Thus, the commodification of the body through the extraction, manipulation, and patenting of genes and cell lines seems to violate intuitions about nature and the person. There is, indeed, a continuing tension between instrumental views of the body as a material object and its social and personal meanings for human beings.

Artists approach this tension against the background of the tensions over the meaning of commodification within their own professional world. As postwar American culture celebrated consumerism, so too Andy Warhol and other American pop artists celebrated the commodity by using icons of fast foods, luxury cars, and movie stars in their art. Commodity culture had both sex appeal and success appeal, but during the 1970s, many artists reacted so vehemently to the commodity nature of art that they sought new venues for their work. For example, earthwork artists, Michael Heizer, James Turrell, and Walter de Maria, moved their art out of the galleries into the desert. The Conceptualists, such as Joseph Kosuth, Larry Weiner, and Douglas Hubler, completely dematerialized their art so that no traditional object could be bought or sold.[46]

Since the 1980s, commercial values have returned to the visual arts with a striking vengeance. Some insiders claim that the market is outweighing the historical significance of art.[47] Artists themselves approach the consumer culture in various and cunning ways. Barbara Kruger's agit-prop productions of photographs, installations, public art, and coffee cups are critical of consumer culture, but at the same time, she defends her own creation of saleable objects. Functioning within the marketplace, she writes, "I began to understand that outside the market there is nothing—not a piece of lint, a cardigan, a coffee table, a human being."[48] Other prominent 1980s artists, including Jeff Koons, Ashley Bickerton, and Haim Steinbeck, were part of a movement labeled "commodity art." Using the signs and symbols of the commodity, these artists produced works that glorified the banality of kitsch. Artists today, in the opening decade of the 21st century, entertain ambivalent responses to the commercialization of their work.[49] Some critique the institutional practices of art consumption as an insiders' game. Others debunk

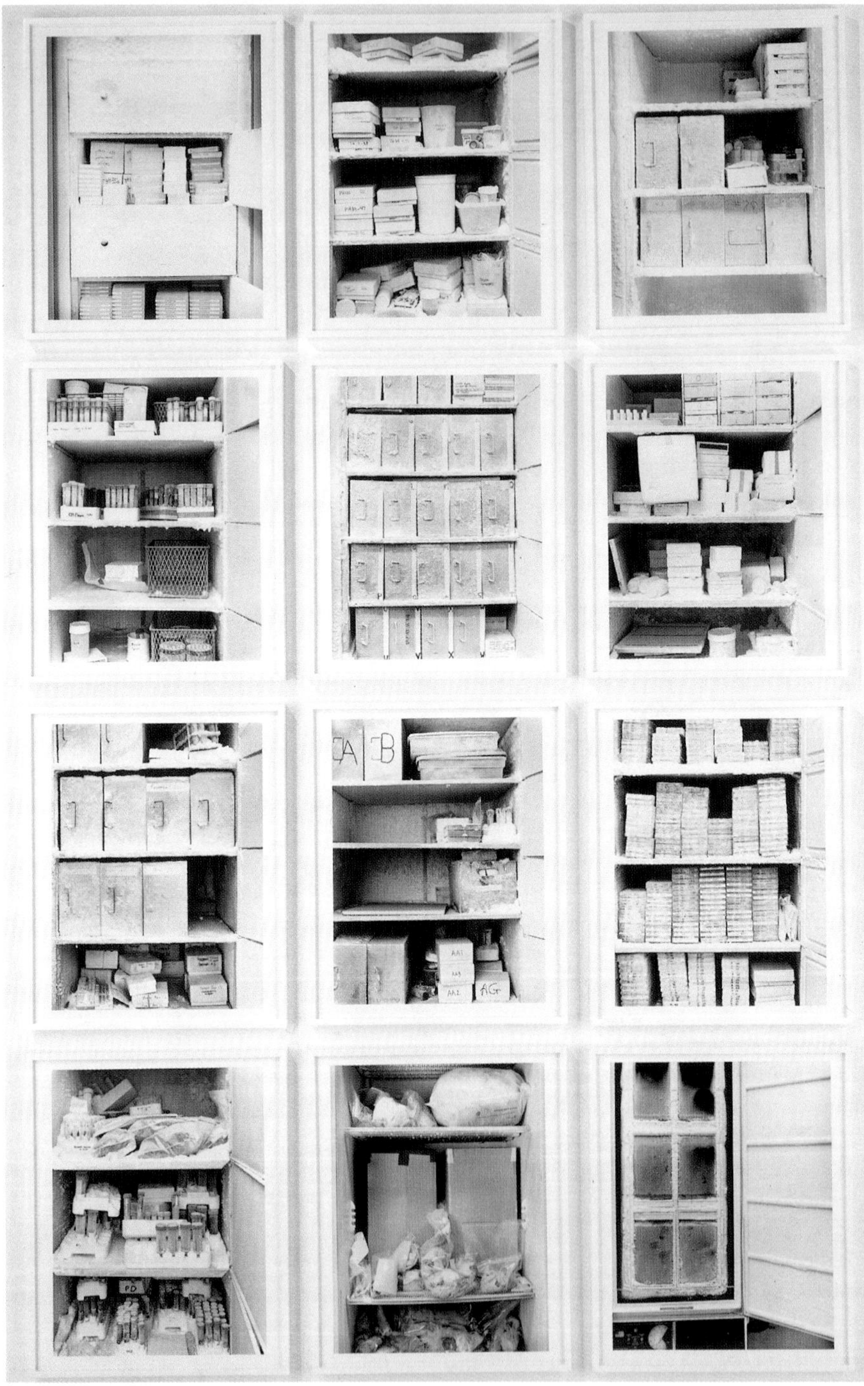

20 Catherine Wagner, *–86 Degree Freezers: 12 areas of concern and crisis,* 1995.

"art as entertainment," calling it a way to pander to the public by making only "feeble demands"on interpretive skills.[50] To those artists who draw visual metaphors from the entire molecular enterprise, commercialization in science is, in part, a means to open discussions about comparative value within the world of art.

Some contemporary artists, like Chrissy Conant and Helen Storey, are pragmatic and bemused—even celebratory—as they draw their images from the state of science and its commercial direction. Others, like Larry Miller and Bryan Crockett, follow the Kantian conviction that living beings should not be used as products or instruments, or as means to an end.[51] Their varied perspectives reflect the broader public ambivalence about biotechnology's unraveling applications. Is body tissue refuse that is freely available as raw material for commercial products, or does it have inherent value as part of a living being? What exactly distinguishes nature from artifact? Are genes the essence of an individual and a sacred part of the human inheritance? Or are they, as Kari Stefansson implied, a form of currency?

Frank Moore's painting, *Oz*, reflects the agonizing complexity of these questions. On the one hand, his work glorifies the worth of genetic research by embedding the products of the genetic revolution in a pot of gold, but at the same time, the chaos in his paintings suggests that science as a commercial enterprise is out of touch with the larger system of ecology and the deeper interlocking values of life.

Notes

1. Interview between Thomas Woodruff and Frank Moore. 2002. The scientific odyssey of an artist. *Update N. Y. Acad. of Sci. Mag.*, April/May, pp. 6–7, 10.
2. For description of these and other ventures, see Lori Andrews and Dorothy Nelkin. *Body bazaar: The market for human tissue in the biotechnology age.* Crown Books, New York, 2001.
3. Stefansson, K. Interview. In *Technology Review* 55, April 2001.
4. Emily Martin. 1990. Science and women's bodies: Forms of anthropological knowledge. In *Body politics: Women and the discourses of science* (ed. M. Jacobus et al), p. 73. Routledge, New York. Also see Leslie A. Sharp. 2000. The commodification of the body and its parts. *Annu. Rev. of Anthropol.*, 29: 287–328.
5. David Magnus, Arthur Caplan, and Glen McGee, eds. *Who owns life?* Prometheus Books, New York, 2002.
6. For the implications of these changes for science, see Sheldon Krimsky. 1999. The profit of scientific discovery and its normative implications. *Chicago-Kent Law Rev.* 75: 15 ff.
7. In 2001, molecular biologist Stuart Newman and biotechnology critic Jeremy Rifkin sought to challenge patent practices by applying for a patent on a human chimera created by mixing human embryo cells with those of a nonhuman animal. They believed that it would be legally difficult for the U.S. Patents and Trademarks Office (PTO) to exclude their chimera without also excluding all other transgenic animals involving human genes. Thus, the PTO might be forced to amend its rules on human/animal patents. The application was rejected because it did not clearly specify the percentage of the chimera that would be human. But it raised an impor-

tant challenge: where should the line be drawn to exclude an invention as being too human to patent?

8. Rebecca Eisenberg, "How can you patent genes?" In Magnus, op. cit. The U.S. Constitution includes a clause that established the foundation of subsequent patent practices: "Congress shall have the power to promote the progress of science and useful arts by serving for limited time to authors and inventors the exclusive right to their respective writings and discoveries." (35US Section 101). Two years later, the U.S. Patent Act provided for patents to be issued to any person who has "invented or discovered any new and useful art, machine, manufacture, or composition of matter, or any improvement of these not before known or used."
9. Arthur Danto. *The transformation of the commonplace.* Harvard University Press, Cambridge, 1983.
10. It is interesting to note that *Fountain* was not included in the Society for Independent Artists exhibition to which it was submitted. It was not understood by the exhibition committee. Steiglitz photographed the original and then the original was lost and had to be replaced by replicas.
11. Martin, op. cit. in Jacobus, p. 73.
12. Kynaston McShine. Andy Warhol Retrospective, MOMA Catalog, 1989. See Benjamin Buchloh. Andy Warhol's one dimensional art, 1956-1966, p. 40.
13. Ibid, p. 40.
14. At this time in his career, Warhol was a commercial artist, directing a New York City advertising agency.
15. Nonorganic examples of body models include the "Glasender Mensch" or Transparent man. He was first placed on display in 1926 at the German Hygiene Museum in Dresden and over the next few decades was rented out for displays throughout the world. This process was described as a new art form, an attempt to display the body artistically in the tradition of Durer and Leonardo. But it was also described in terms of its commercial value and associated with the devaluation of human beings in the Weimar Republic. Klaus Vogel. 1999. The transparent man—Some comments on the history of a symbol. In *Manifesting medicine: Bodies and machines* (ed. Robert Bud, B. Finn and H. Trischler). Taylor and Francis, London.
16. Korperwelten. (Body Worlds), Catalog for exhibition at Mannheim, Germany, 1998. Also, see Mary Roach, "Stiff: The Curious Lives of Human Cadavers." W.W. Norton, London and New York, 2003, pp. 286–290.
17. National Library of Medicine Exhibit, "Dream Anatomy," Michael Sappol, Curator, Catalog, October 9, 2002–July 31, 2003. http://www.nlm.nih.gov/exhibition/dreamanatomy.
18. Incarnate. Exhibit at Gagosian Gallery, New York. Catalog, Gagosian, New York, February 1998.
19. Body materials have often been used in visual art. Manzoni, for example, used feces, canned into a commodified exchangeable good, as a comment on the fetishization of art and middle class propriety. Chris Burden, Vito Acconci, and Dennis Oppenheim employed the body as sculptural material during the 1970s. Their actions were reactions against art as fungible goods. More recently, Orlan and Stersak have used blood and body tissue in their work. And hair is a frequent medium through which artists such as Mona Hatoum explore the integrity of the body and its politics. William Delvoye created *Cloaca,* a sculpture simulating the digestive system that emits chemically produced feces. He then sells the freeze-dried vacuum-packed feces as "cybershit." The cost? $1500. Advertisement in *Flash Art* July-September, 2002.

20. Quoted in *The Independent*, April 14, 2002, and on her website.
21. John Weeks. 1989. Art pair fined over foetus earrings. *Daily Telegraph* (London) February 10, p. 3.
22. Jacquelin Urla and Alan Swedland. 1995. The anthropometry of Barbie. In *Deviant bodies* (ed. Jennifer Terry and Jacqueline Urla), pp. 305–306. University of Indiana Press, Bloomington. See also, M.G. Lord. *Forever Barbie: The unauthorized biography of a real doll.* Morrow, New York, 1994
23. Catalog of the touring exhibition, "Primitive Streak", beginning at London's Institute of Contemporary Art and ending in New York City at the World Financial Center, 1997.
24. Harvard has since obtained patents for the mouse in Japan and much of Europe, but in 2002 the Supreme Court of Canada denied a patent arguing that higher forms of life do not fit the definition of an invention.
25. Donna Haraway describes the oncomouse as "a fleshy mystery in a secularized salvation history of... scientific knowledge, progress, democracy and economic power." Donna Haraway. *Modest witness, second millenium: Female man meets oncomouse.* Routledge, New York, 1996.
26. Interviewed in Marvin Heiferman and Carole Kismaric. Picturing the genetic revolution. Catalog for Paradise Now Exhibition, Tang Museum, 2001.
27. At a recent public appearance by James Watson, at CUNY Graduate Center in NYC, in conjunction with a festival of visual arts exhibits, DNAge (2003), Dr. Watson stated that DNA was a new theology and like all theologies, it spawns its own kind of art.
28. Lehman Maupin Gallery, February 20 to April 6, 2002.
29. Lawrence Osborne. "Fuzzy Little Test Tubes." *The New York Times Magazine*, July 30, 2000, pp. 40–45.
30. Mark Jagels. 2002. Dr Moreau has left the island: Dealing with human animal patents in the 21st century. *23 Thomas Jefferson Law Rev.* pp. 115–147.
31. Gary Taubes. 2002. Anti-body drug revival. *Technol. Rev.* July/August, p. 65.
32. www.ppl-therapeutics.com/welcome/welcome/html
33. David Bowers. 2000. Will Frankenfood feed the world. *Time* 3, July, pp. 48–49.
34. "Bald chicken needs no plucking," BBC News, May 21, 2001.
35. Robert F. Service. 2002. Mammalian cells spin a spidery new yarn. *Science* 295: 419–421.
36. This is significantly different from European laws which are based on different principles giving priority to questions of human dignity ("ordre public"). European countries have not allowed DNA patenting. See E. Richard Gold and Alain Gallochat. 2001. The European biotech directive: Past as prologue. *European Law J.* 7: 331–366.
37. U.S. Patent number 6,211,429, April 3, 2001.
38. Margaret Radin. *Contested commodities.* Harvard University Press, Cambridge, 1996.
39. John Moore v. Regents of the University of California, 793 P 2nd 479 (Cal 1990).
40. James Meek. 2000. Poet attempts the ultimate in self-invention: Patenting her own genes. *Guardian* (London) February 29, p. 3.
41. www.dnacopyright.com
42. Harriet A. Washington. 1994. Henrietta Lacks — An unsung hero. *Emerge* October, p. 29. See also R. Skoot. 2000. Henrietta's dance. *Johns Hopkins Mag.* April.
43. Mike Mitka. 2002. Banking on genes. *J. Am. Med. Assoc.* 288: 2951–2952.
44. www.decodegenetics.com

45. Anthony Synnott. *The body social.* Routledge, London, 1993.
46. Lucy Lippard. *Changing: Essays in art criticism.* E.P. Dutton, New York, 1971.
47. Art dealer Jeffrey Deitch comments that the 1980s will be remembered more for the exchange value placed on artworks than for the artworks themselves.
48. Barbara Kruger quoted in Carol Squiers. 1987. Diversionary (Syn)tactics. *Art News* February.
49. See discussion by Irving Sandler. *Art of the postmodern era.* Westview, Boulder, 1998.
50. Clement Greenberg. 1978. Avant garde and kitsch. In *Art and culture: Critical essays.* Beacon Press, Boston.
51. See the special issue, Who's afraid of commodification. *Kennedy Institute Ethics J.* vol 11, 3, September 2001.

Science As Culture: Through the Artist's Lens

THE SCIENCE OF GENETICS has rapidly become a dominant cultural narrative complete with its own superstars and corporate megabucks. Amplified by the media machine, this field of systematic inquiry brings into focus concerns over future definitions of "life." Since the late 1980s, a stream of visual artists have entered this cultural discourse, initially through identity politics in the form of "body art" and more recently through molecular models.[1] Their work comments on issues of identity, normality, and authenticity raised by the daunting new imaging technologies and the possibilities of germ-line transformation. What are the implications of this powerful discipline for the meaning of personhood, the integrity of individuals, and future control over human reproduction?

Our findings in *The Molecular Gaze* reinforce the notion that science has left the private confines of the laboratory and become part and parcel of the social order. Through the artist's lens, science is not a dry mechanistic set of equations but an enlivening interrogation of the malleable dimension of life. By interweaving concerns in the biosciences with genres in art and its history, we illustrate some of the underlying motif structures engaging culture over time. This may in considerable part be due to the fact that DNA itself is such an elastic molecule, a pliable yet stable entity that can be pushed, pulled, mixed and rematched, reconfigured and morphed.

Building on the iconography of the genetic code—the double helix, the chromosome, and the autoradiograph—artists have appropriated these notational images, picturing DNA in various guises. Some follow the tradition of abstract art, using calligraphic marks and symbol sets. Others incorporate molecular imagery into the historical genre of portraiture. Still others use these signs to critique the reductionist premises intrinsic to this scientific field and to question its relevance for understanding the genuine complexities haunting human life.[2]

Emerging as the "new grotesque," a revival of monsters, mutations, and the abject in art, figures with bodily aberrations or congenital deformations portray the body as an anxious subject. In an age of volatile extremes, the body has become another realm of fantasy. In addition to harboring deep-seated religious and moral ambivalence about the flesh, the very concept of mutation evokes the fear of a future "gruesome parade of horribles." Images of these horribles are appearing in galleries and museums of art.

The classic concept of the chimera has reemerged as a formidable creature in contemporary art with figures that cross species boundaries and defy

Linnaeus's taxonomic order. Some refer to actual transgenic experiments: For others, transgenics means transgression, metaphorically alluding to the manifest tensions in the postmodern world. The blurring of social boundaries, the disruption of traditional definitions, the threats to authenticity and species integrity are pivotal to the polemics associated with this molecule.[3]

The problematic history of eugenics has resurfaced at a juncture where assisted reproductive technologies converge with techniques of cloning and embryo manipulation. These technologies create scientific capabilities for the selection of traits and even for the "breeding of better people." Many artists are subsequently drawn to the questions of selected breeding and family lineage.

The commercial potency of genetics for the development of profitable diagnostic techniques and therapeutic products has embedded this science deep in corporate culture. Genes and cells have effectively become commodities. The images that explore the implications of patenting genes and cell lines employ legal certificates, patent application graphics, advertising logos, and mock corporations to make their point. Reducing the body to a bio-archive that is mined and harvested, patented and licensed, raises serious questions addressing life's sanctity.

Sci-art: Emergence, Convergence, Divergence

Reviewing the work of hundreds of artists responding to genetics, documenting their "molecular gaze," we have observed some revealing changes taking place in the relationship between art and science over the past few decades. The early modern relationship between art and science developed on a model which assumed that science, representing progress and truth, was the ideal way of investigating and classifying the world. Today, however, the relationship is less idealistic, less optimistic, as many artists question the essentialist and prevailing assumptions of science. Much of their work tilts toward the myriad applications of genetics—the possibilities of engineering human, animal, and plant life—for their implications affecting individuals, families, our concepts of nature, and our altered reproductive and evolutionary future. And many of these images challenge traditional notions of beauty, significance, and order.

The contemporary art included in this text points to several questions about the complex and changing narrative between art and science. How is this relationship altered by the ubiquitous use of instrumentalized visualization techniques? What is the meaning of veracity, authenticity, and aesthetics in these two domains? How do the images emerging from both science and art relate to the contours and concerns of their distinct professions? What is the effect of recent cultural fascination with science in general and genetics in particular?

Science and art have been radically influenced by new techniques and increasingly sophisticated instruments. Genetic scientists such as Craig Venter employ computerized gene sequencing technologies to fabricate chimeras in their labs. Dr. Venter has recently discussed his plans to develop "the first artificial life form in a laboratory dish."[4] In various parallel ways, artists employ Photoshop and its allied software programs to create heterogeneous images through the operation of cut-and-paste techniques in film, photography, sculpture, and emergent forms of digital art. Rapid prototyping, animations, and "special effects" bring into being extraordinary visual results by turning forms into numbers, and numbers back into forms. Most recently, researchers Vladmir Mironov and Thomas Boland are experimenting with rapid prototype technology in an effort to one day "print" human organs. This technique, known as 3d bioprinting, replaces standard ink cartridges with live cells in suspension.[5] Goaded by heuristic experimentation and novel generation of abstract modeling, art and scientific practice currently share a similar and overlapping technological base.[6]

Visual instruments have changed scientific procedures in ways that influence these values. The state of the body is read through tests—the PSA antigen test, magnetic resonance imaging, the ultrasound, X-ray, and new microscopy techniques that operate at nanoscales. Life on the screen is a reality for science, and measurement is often more important than direct observation. Scientists represent their data produced through sophisticated optical instruments in the form of graphs, charts, formulas, and photographic images. Like artists, they manipulate, alter, and edit these images to contain complexity and emphasize characteristics that effectively illustrate their interpretation of the data.[7] Their "inscription devices," writes Bruno Latour, create "audio-visual spectacles" intended as sources of information, but also as tools of persuasion.[8] They are compelled to convey information in a comprehensible and attractive manner that will garner attention and persuade the viewer—the scientist, the patient, or the public—of their accuracy and authenticity.

Photo-based images have long raised questions of veracity. Mechanical means of reproduction and instrumentalized visualization techniques increase the capacity to augment a reality, or reduce or enhance an iconic resemblance, and even to lie. Scientific photographs are edited so as to articulate a visual proposition. This method may be helpful in conveying information, but, in a historical and epistemological sense, manipulation of scientific artifacts is questionable if not devious. Observing the dramatic increase in the role of images in science over the past 20 years, Julio M. Ottino, a chemical engineer, calls for a system of guidelines specifically tailored to scientific image manipulation and enhancement to make these images more accountable in a scientific sense.[9] Similarly, historian of science Peter Galison writes about the powerful ambivalence about images within the sciences: "We must have images: we cannot have images." Pictures are important aids to conceptual thinking in science, yet scientists oscillate in their use of images—"torn between the concrete and the abstract."[10]

The manipulation of images has implications for related values of authenticity. Geneticists, by definition, are searching for the underlying sources of pathology or mutation, but they are also challenging notions of authenticity as they turn nature into artifact by altering the genes of tomatoes, corn, goats, sheep, and on and on. If something looks too good to be true, has it been tampered with?

Authenticity is also a fundamental position for the artist. It is no small point for photographers of a certain stripe, as one representative example, to define the natural or abnormal, the healthy or diseased, by way of documenting body anomalies. In addition, a rather novel subgenre of portaiture has recently emerged employing the nucleotide sequences of its subjects in pursuit of obtaining their embodied essence. Artists of quite another stripe explore the issue of authenticity by addressing the possible range of representation of ambiguous transgenic creatures.

On Beauty in Science and Art

Often scientists enhance scientific images to increase their aesthetic values. Indeed, when discussing their work, scientists often describe their images, theories, and experiments as "beautiful," "aesthetic," or "elegant." James Watson described DNA as the "prettiest way for a polynucleotide chain to fold up."[11] Some critics, such as Denna Jones, write about "a symbiotic relationship between truth and beauty which makes science and art such appropriate bedfellows."[12] But the issue of beauty itself has become a slippery slope in both art and science. The correlation between truth and beauty is grounded in classical aesthetic theory. From the physiognomy studies of the 19th century to the current assumptions underlying racial profiling, truth has been correlated with idealized physical attributes. This is based on a false premise that has often led to punitive social classifications.

Beauty is frequently associated with moral implications.[13] Elaine Scarry, in her book *On Beauty and Being Just,* makes a case for "unbanishing" beauty from studies in the humanities. Her description of beauty as a form of clarity can be taken as a stance on the aesthetic/ethical conjugation while avoiding its numerous fundamental dilemmas. Whereas beauty as a form of cognitive recognition was a quality used by early humans for survival, its application in a culture where appearances can evoke multiple judgments is questionable. Ronald Jones's sculpture (Fig. 1) is a case in point. Appropriating Brancusi's style of modernism (Fig. 2),[14] Jones fashions a sculpture in bronze, wood, graphite, and steel. On first glance, the viewer confronts an elegant abstract sculpture, refined and attuned to its material nature. But beauty here acts as a decoy, thwarting the viewer's expectation.

When the viewer in fact is informed of the title of the piece, *Untitled (Core of the Human Retrovirus)* (1989), mutual associations between beauty and the good dissolve. In a world of nonretinal art and hidden genetic engineering, visual information may not yield what was once a more primary sense—visuality as discernment.

The enchanting beauty of many scientific photographs, diagrams, and models has entitled us to explore the comparative differences between the role of aesthetics in visual art and in science. Scientific images can be truly beautiful, but their beauty may very well have little to do with aesthetics as professionally defined. Contemporary philosophers of aesthetics and historians of art, such as Nelson Goodman, Ernst Gombrich, and Jurgen Habermas, are less concerned with what is beautiful than with how the aesthetic value of a work is culturally and perceptually defined.

In science, beauty is a subset, a by-product, not a primary objective. It is secondary to the goal of precision and communication. Intending to document and elucidate, scientific images, writes historian Sian Ede, "are largely devoid of any artists' imaginative interpretation."[15] Art, on the other hand, is *not necessarily beautiful.* Indeed, many of the artworks that appear in this book, the work of Witkin, Bellmer, and Sherman, for instance, can hardly be called beautiful in a traditional sense. Rather, they are radical rejections of the inauthentic in a standardized society where beautiful images are often constructed and engineered to portray an idealized version of self. These images are instead reflections of anxiety, animosity, even disgust; and in many cases, they are intended to shock.

Ultimately, the images generated by scientists and those provided by artists are based on quite different epistemologies.[16] Science and art at a given time and place may share a cultural context and both draw referents from the same cultural milieu. Yet, they represent quite distinct ways of knowing the world. Whereas the discipline of science aspires to detachment and objectivity, the practice of art is embedded in personal and emotional interpretations. Yet, even sometimes there are acknowledged or accidental "crossing-overs."

The intentional detachment of scientific images reflects the fact that science is evidence-based. Depicting the results of an experiment or probe, their function is to express accurately and unambiguously a proposition in science or a clinical condition with a precision that leaves minimum room for subjective interpretation.[17] Moreover, scientific images will necessarily change with the advance of knowledge. Truth in science is temporal, based on evolving facts as they are known at any given time and accepted by the collectivity—the consensus within the scientific community. In contrast, a good deal of art thrives on ambiguity or aspires to subjectivity, to a truth based less on accepted fact than on personal perceptions and emotional resonance. A work of art, therefore, is hardly temporal. One surely does not alter a Giotto fresco because our knowledge about anatomy has advanced.[18]

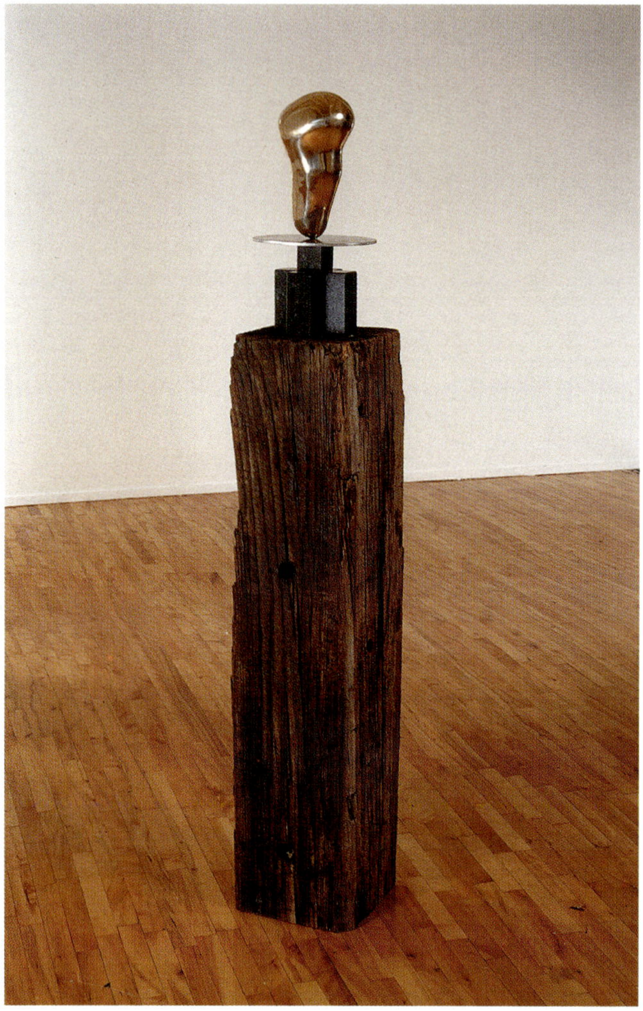

1 Ronald Jones, *Untitled (Core of the Human Retrovirus)*, 1989. (Human T-lymphotic virus Type I which contains protein p25, the RNA which carries the virus's genetic information, and the enzyme reverse transcriptase, corresponding to the viral RNA.)

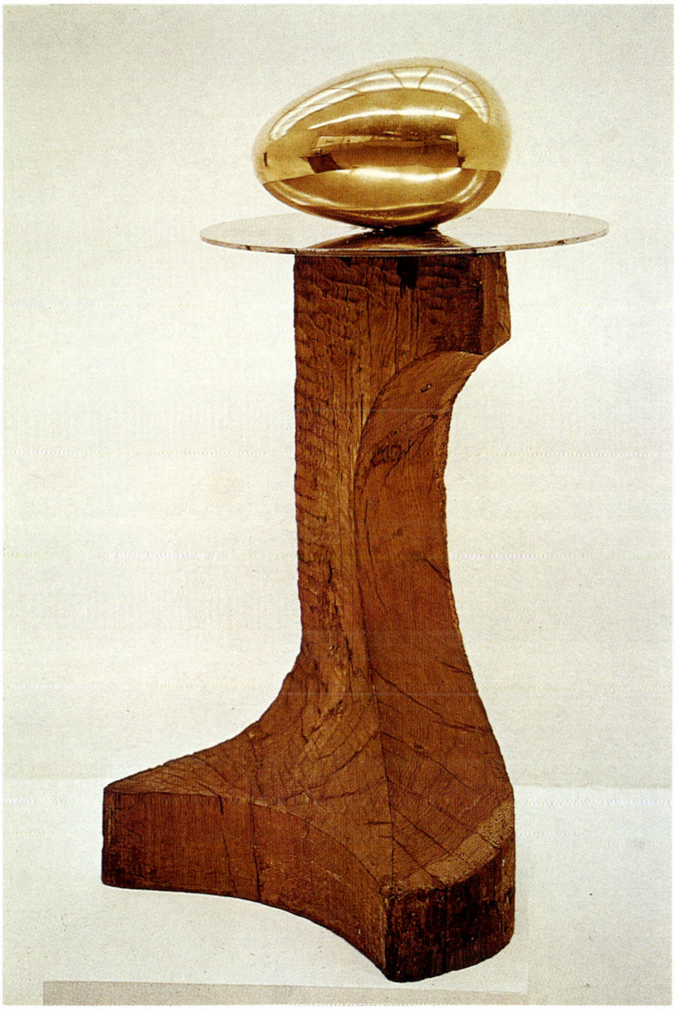

2 Constantin Brancusi, *The Origin of the World*, 1924.

Toward the Future: Animating Life

Artists, ever attracted to new aesthetic strategies, tools, and materials, often discover them in the scientific laboratory. Many artists are embracing the sophisticated optical instruments and photographic techniques that scientists use to visualize the cells and tissues of the body. Ovalle, Clarke, Tsiaras, Rees, and Kac, among others, deploy computers, high-speed cameras, MRIs, and other high-tech instruments to create their images. As they borrow new techniques from the armamentaria of science, this blurs the distinction between scientific images and art. Indeed, in some cases, the distinction between art and science becomes a matter of intent or context.

Just as scientists build on the theories and concepts of their mentors, so artists position their work in the context of historical trends and rivalries within their own profession. Although contemporary artists often seem to be developing novel approaches, they are building on archetypal images and methods that have been important throughout the entire history of art. Their figures include repeated references to certain historically meaningful images—the chimera, the cyclops, the monster, the classic torso. In addition, their use of abstract signs has historical resonance with the role that language and, more recently, semiotics have played in the history of the visual arts.

Artists, in their approach to science, are also challenging the historical boundaries of their own field. In some cases, artists' perspectives seem to reflect formal if not professional concerns. Their images of commercialization in science, for example, are a means to comment on the dilemmas and intriguing disputes over commodification in their own profession. But is commodification—both in art and in science—an inevitable force? Or is it the result of mass media-induced desire? When artists such as Kac and Orlan transgress traditional expectations by using living matter as mediums in their creations, are they questioning practices in science and art, or just engaging in yet another commercial spectacle?

Some of the images that appear in these chapters are celebrations of science, others express ambivalence about biotechnology and especially about those applications that seem to accelerate conformity and standardization. They visually express the mixed reactions that are frequently voiced in the public response to many aspects of genetics—the mix of awe and skepticism, of hope and fear. Is this science a source of salvation or a means of exploitation? Will its promises exceed its perils? Will genetic discoveries become a boon to health or only a boon to scientific careers at an unknown cost to future generations?

Despite the persistence of such ambivalence, science in the 21st century is in vogue. Physics and mathematics are in fashion in the theater, where plays about Einstein, Neils Bohr, and John Nash are blockbuster hits. Fashion models pose on the cover of a new glossy science magazine, *Seed*, that associates science with glamour and cool. Corporate ads integrate science with the beautiful people. Science has enjoyed a virtual renaissance in books authored by scientist-turned-popular writers, and in extravagant televised documentaries, infomercials, and media reports.

So too, genetics, especially, is under the spell of this "lab coat chic."[19] The musical, literary, and performing arts have responded to genetic models and hereditarian themes as well. Software programs now turn DNA sequences into musical scores. Susan Alexjander composes music derived from the light absorption spectra of the four nucleotide bases. They are programmed and synthesized with vocals, cello, tabla, and violin to produce Zen-like sounds.[20] Two Spanish microbiologists, Aurora Sanchez-Souza and Fernando Baquero, are composing "genome music," basing this music on the chemicals that make up the DNA molecule. They have turned the nucleotide sequence of several microbe genes into compositions.[21] The melodies, determined by the genome's basic sound, resemble Gregorian chants. And in the other direction, Frank Gillette, in his digital and visual composition, *The Broken Code (for Luria)* (2002) (Fig. 3), turns a Gregorian chant into a dance of mitosis.

Jonathan Tolins's play, *The Twilight of the Golds* (1993), builds on the dilemmas of genetic testing by drawing on the responses from a family who discovers that their unborn child has the "gay gene." A popular Hollywood film realized in 1997, *GATTACA*, turns on the bureaucratic deployment of

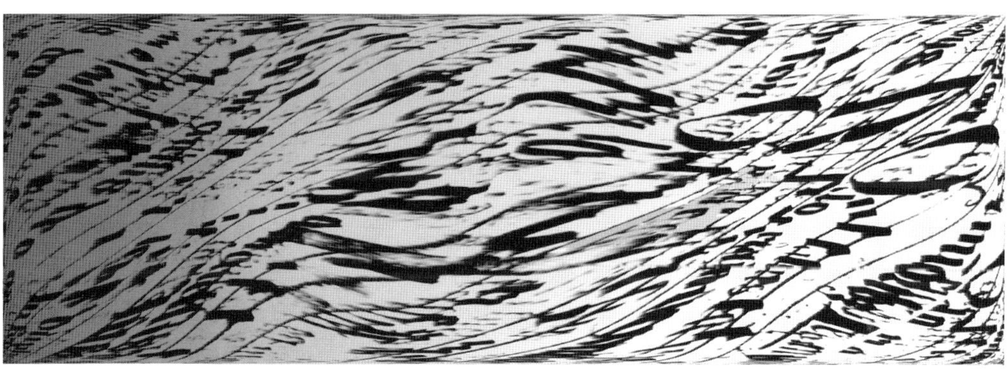

3 Frank Gillette, *The Broken Code (for Luria)*, 2002.

body fluids and DNA that purportedly determines everything about a person. A genetically perfect elite, conceived in the laboratory, reigns over the genetically "invalid" conceived in the usual way. And cloning is a ravenous subject for science fiction narratives and mass culture films.

The appearance of genetic signs and symbols in the visual arts is part of this cultural trend. Nucleotide sequences and double helices appear in art galleries and museums as well as in science illustrations. Mutants and monsters appear in art books as well as in comics. Genetics has become a spectacle, a new mythology, even a theology. A source of power and progress, the helix has a "double edge."[22] Newspaper critic Dennis Overbye describes the ambiguity inherent in this spectacular science. Science, he writes, has given us "godlike powers and responsibilities... and with it the possibility of truly godlike failures."[23] The molecular gaze of artists reflects this ambivalence, replacing the modernist image of science as an unquestioned source of progress with an open set of questions and critical reflections on the intricate ambiguities of our genetic age.

What then can we expect to be the effects of the "molecular gaze," as artists enter the cultural discourse about this science and its pervasive influence? Do DNA images advance the public understanding of genetics and increase sensitivity to its social and ethical implications? The answers are elusive. Visual art can help to explicate and popularize an often obscure science, translating difficult information and illustrating complex social issues in persuasive and culturally meaningful ways, but artists cannot really communicate concepts in science that are beyond—and even defy—common understanding. Rather, their work reflects their individual responses and social concerns. Moreover, art has its own deliberate obscurities and complexities that follow from its unique historical and symbolic associations. There may be better ways to communicate and popularize science.

Encouraged by showmen such as von Hagens and extremists such as Orlan, "DNA art" could be dismissed as merely an illustration of scientific precepts, or as a relic of the 19th-century preoccupation with freaks, or simply another means of mass culture entertainment. Contextualizing the genetic revolution in an art gallery or museum could merely foster the pervasive tendency to eventually transform every new entity into consumer culture kitsch, but there remains little doubt that the vision of artists can contribute to our understanding of the real ambiguities of a powerful science—a science that is having a pervasive, saturating, and perplexing influence on our self-perceptions, social and political practices, as well as our cultural beliefs.

Notes

1. Jeff Rian. 1993. What's all this body art? *Flash Art* Jan/Feb., pp. 50–53.
2. Daniel C. Dennett. 2003. The mythical threat of genetic determinism. *Chronicle of Higher Education* January 31.

3. Suzanne Anker. 2003. The double helix comes of age: Artists explore DNA. *Seed Magazine* March/April, pp. 90–99.
4. Jeremy Rifkin. 2003. Dazzled by the science. *The Guardian* January 14.
5. Printing body parts. *3D Bioprinting*, January 30, 2003. At http://www.cadcamnet.com.
6. Stephen Wilson. *Information arts: Intersections of art, science and technology.* MIT Press, Cambridge, 2002.
7. Michael Lynch and Steven Woolgar. *Representation in scientific practice.* MIT Press, Cambridge, 1990.
8. Bruno Latour. *Science in action.* Harvard University Press, Cambridge. 1987, pp. 68–71.
9. Julio M. Ottino. 2003. Is a picture worth 1000 words? *Nature* 421: 474–476. Also see Felice Frankel. *Envisioning science: The design and craft of the science image.* MIT Press, Cambridge, 2002.
10. Peter Galison. 2002. Images scatter into data: Data gather into images. In *Iconoclash: Beyond the image wars in science, religion and art* (ed. B. Latour and P. Weibel), pp. 300–323, MIT Press, Cambridge.
11. James Watson. *The double helix.* Atheneum, New York, 1968.
12. Denna Jones. *Working drafts.* Wellcome Trust, London, 2002.
13. Elaine Scarry. *On beauty and being just.* Princeton University Press, New Jersey 1999.
14. Suzanne Anker. "Gene Culture: Molecular Metaphor in Visual Art." *Leonardo.*
15. Sian Ede, ed. *Strange and charmed: Science and the contemporary visual arts.* The Calouste Gulbenkian Foundation, London, 2000.
16. Leo Steinberg. 1986. Art and science: Do they need to be yoked? *Daedalus* Summer, pp. 1–16.
17. Art historian James Elkins, however, writes that art historical methods are valuable in analyzing all images. For they "possess the most exact and developed language for the interpretation of pictures."
18. Conversations with Frank Gillette.
19. Dennis Overbye. 2003. Lab coat chic: The arts embrace science. *The New York Times* January 28, p.1, Science Times.
20. Susan Alexjander. Sequencia. (CD Recording.) Science and the Arts, Berkeley, 1994. Also see New Music composer Patrick Grant's CD entitled *Genome: The Autobiography of a Species in 23 Movements* (2003) based on Matt Ridley's book, *Genome: The Autobiography of a Species in 23 Chapters* (HarperCollins, New York), 2000. On another note, see Sirasak Teparkum's CD *Genomic Music* based on differing gene sequences in diverse organisms such as cat, dog, jasmine, rice, and orchids.
21. Random Samples: 2002. *Science* 298: 2325.
22. This is the metaphor used by several writers, e.g., Joseph Alper et al., ed. *The double-edged helix: Social implications of genetics in a diverse society.* Johns Hopkins University Press, Baltimore, 2002.
23. Overbye, op. cit.

List of Artists and Their Works

Anderson, Laurie. *What You Mean We?*, 1986. Scan of video still. Courtesy of the artist and Canal St. Communications, p. 130.

Anker, Suzanne. *Material Powers*, 1999. Digital print; size variable. Courtesy of the artist, p.113.

Anker, Suzanne. *Micro Glyph (Soma Font)*, 2000. Silkscreen and acrylic on wooden panel; 4 x 6 feet. Collection: Goldblum Lévy, Switzerland, p. 29.

Anker, Suzanne. *Zoosemiotics: Primates, Frog, Gazelle, Fish* [detail], 1993. Glass vessel, water, steel, hydrocal, metallic pigment; Installation: 3.7 cm x 8.2 cm x 1.52 m. Courtesy Universal Concepts Unlimited, New York, p. 30.

Ashbaugh, Dennis. *Designer Gene*, 1992. Mixed media on canvas; 185.4 x 274.3 cm. Courtesy of the artist, p. 32.

Aziz + Cucher. *Chimera #1*, 1998. Color photograph; 60 x 30 inches. Courtesy of the artists and Henry Urbach Architecture, New York, p. 108.

Aziz + Cucher. *Chimera #2*, 1998. Color photograph; 60 x 30 inches. Courtesy of the artists and Henry Urbach Architecture, New York, p. 108.

Barney, Matthew. *CREMASTER 4*, 1994. Production photograph. Courtesy Barbara Gladstone Gallery, New York, p. 99.

Bellmer, Hans. *La Poupée (The Doll)* [detail], ca. 1937–1938. Photograph and drawing on paper; 184 x 648 mm., p. 61.

Boltanski, Christian. *The Detective* [detail], 1972–1973. 408 photographs in various sizes. Courtesy of The Art Institute of Chicago, p. 15.

Borland, Christine. *A Treasury of Human Inheritance, Huntington's Disease, Entres' Case*, 2001. Silver, steel, and agates; large mobile with 124 agates; 260 x 180 cm. Courtesy Sean Kelly Gallery, New York, p. 138.

Brancusi, Constantin. *The Origin of the World*, 1924. Polished Bronze; 19 x 28.5 x 17.5 cm. Collection: Musee National d'Art Moderne, Centre Georges Pompidou, Paris, France, p. 191.

Buck, Chris. *Dolly*, 1997. Chromogenic print; 30 x 30 inches. Courtesy of the artist, p. 128.

Burson, Nancy. *Untitled (Jessica and her Mom)*, 1991. Gelatin silver print; 15 x 15 inches. Courtesy of the artist, p. 68.

Chadwick, Helen. *Nebula*, 1996. Cibachrome photographs, perspex; 18 x 2 inches, each of 4 circular panels, 23 x 2 inches, each of 2 circular panels; 1 oval panel 35 x 2 inches. Courtesy Zelda Cheatle Gallery, London, p. 147.

Chalmers, Catherine. *Rhino*, 2000. C-print; 30 x 40 inches. Courtesy of the artist, p. 168.

Chapman, Jake and Dinos. *Tragic Anatomies* [detail], 1996. Fiberglass, resin, paint, smoke devices; dimensions variable. Courtesy Jay Jopling/White Cube, London, p. 47.

Chapman, Jake and Dinos. *Zygotic acceleration, Biogenetic de-sublimated libidinal model (enlarged x 1000)*, 1995. Mixed media; 59 x 71 x 55 inches; 150 x 180 x 140 cm. Courtesy Jay Jopling/White Cube, London, p. 65.

Clarke, Kevin, *Portrait of James D. Watson* [panel 5], 1998–1999. Color print under acrylic panel; 73.7 x 200.6 cm. Courtesy of the artist, p. 34.

Clarke, Kevin. *Portrait of Jeff Koons*, 1993. Cibachrome on aluminum; 121.9 x 182.9 cm. Courtesy of the artist, p. 35.

Clemente, Francesco. *Silence*, 2001–2002. Watercolor on paper; 44.75 x 44.25 inches. Courtesy Robert McKeever/Gagosian Gallery, New York, p. 144.

Conant, Chrissy. *Chrissy Caviar*® [detail], 2001–2002. Human eggs, human tubal fluid, liquid silicone, polyethylene, nylon, glass, brass, refrigeration equipment; 52 x 49 x 47 inches. Courtesy of the artist, p. 161.

Cooper, Justine. *Moist*, 2002. Video still. Courtesy of the artist, p. 178.

Cragg, Tony. *Code Noah*, 1988. Bronze; 275 x 100 x 100 cm. Collection: Mr. and Mrs. Ware Travelstead; Courtesy of Marian Goodman Gallery, New York, p. 28.

Crockett, Bryan. *Ecce Homo*, 2000. Marble, epoxy, stainless; 108 x 42 x 42 inches. Collection: JGS, Inc.; Courtesy Lehmann Maupin Gallery, New York, p. 166.

Crockett, Bryan. *Pride, Lust, Gluttony, Sloth, Greed, Envy, Anger*, 2001. Cultured Marble; 7 parts, variable dimensions [edition of 3]. Courtesy Lehmann Maupin Gallery, New York, p. 167.

Dali, Salvador. *Butterfly Landscape, The Great Masturbator in Surrealist Landscape with DNA*, 1957–1958. Collection: Private, p. 24.

Danuser, Hans. *Tableau I 3* from *Frozen Embryo Series I* [4 parts (I 1 - I 4)], 1998–2000. Silver gelatine print; 69 x 55 inches each, mounted on 2 mm aluminum sheet. Courtesy of the artist, p. 143.

Daumier, Honoré-Victorin. *Jacques Lefebvre: The Fine, Trenchant Spirit (L'esprit fin et Tranchant)*, ca. 1832–1835 (cast ca. 1948–1952). Bronze; 7.75 x 4.63 x 5.5 inches; 19.7 x 11.7 x 14.0 cm. Collection: Smithsonian Hirshhorn Museum and Sculpture Garden, Washington, D.C., p. 13.

Daumier, Honoré-Victorin. *The Parisian Janitor (Le Portier Parisien)*, ca. 1840–1862 (cast ca. 1951–1957). Bronze; 6.25 x 2.5 x 3.25 inches. Collection: Smithsonian Hirshhorn Museum and Sculpture Garden, Washington, D.C., p. 13.

Degas, Edgar. *The Little Fourteen-Year-Old Dancer* [left profile], executed ca. 1880 (cast 1922). Bronze, partially tinted, with cotton skirt and satin hair-ribbon, wood base; Height (w/out base) 39 inches; 99.1 cm. Collection: H.O. Havemeyer Collection, Metropolitan Museum of Art, New York, p. 14.

Ernst, Max. *Here Everything Is Still Floating (Le vapeur et le poisson)*, 1920. Cut-and-pasted halftone relief and pencil on halftone relief on paper; 4.13 x 4.9 inches; 10.5 x 12.4 cm. Collection: Museum of Modern Art, New York, p. 88.

Ernst, Max. *Stratified rocks, nature's gift of gneiss lava*, 1920. Gouache and pencil on chromolithograph of anatomical subject on board with ink; 6 x 8.13 inches; 15.2 x 20.6 cm. Collection: Museum of Modern Art, New York., p. 87.

Fry, Benjamin. *Genomic Cartography: Chromosome 21*, 2001. Chromogenic print; 8 x 8 feet. Courtesy of the artist, p. 40.

Fuss, Adam. *Invocation*, 1992. Unique cibachrome photogram; 40 x 30 inches. Courtesy of the artist and Cheim & Read, New York, p. 126.

Gauguin, Paul. *Tehamana Has Many Ancestors (Merahi metua no Tehamana)*, 1893. Oil on canvas; 76.3 x 54.3 cm. Courtesy of The Art Institute of Chicago, p. 20.

Gessert, George. *Streptocarpus hybrid 523*, Hybridized 1997, first bloomed 1998. Mounted cibachromes; 122 leaves, each 3 x 5.25 inches; Flowers 2 inches across. Courtesy of the artist, p. 67.

Gillette, Frank. *A Tangled Bank* [detail], 2000. Archival digital images mounted on aluminum; twelve 45 x 15 inches. Courtesy of the artist and Universal Concepts Unlimited, New York, p. 105.

Gillette, Frank. From the *HCE Series*, 1999–2000. Courtesy of the artist and Universal Concepts Unlimited, New York, p. 105.

Gillette, Frank. *The Broken Code (for Luria)*, 2002. Chronogenic print on aluminum. Courtesy of the artist and Universal Concepts Unlimited, New York, p. 193.

Gormley, Antony. *Bodies in Space IV*, 2001. Forged ball bearings; 206 x 86 x 72 cm. Courtesy Jay Jopling/White Cube, London, p. 41.

Grey, Michael Joaquin. *Carbon Codex 15 x 8*, 1992. Graphite tablet, CNC milled; 61 x 122 x 5 cm. Courtesy bitforms, New York, p. 38.

Grünfeld, Thomas. *Misfit (St. Bernard)*, 1994. Tierpräparation; 60 x 120 x 70 cm. Saatchi Collection, London; Courtesy of the artist, p. 81.

Hausmann, Raoul. *Tatlin at Home*, 1920 (destroyed). Collage of pasted papers and gouache; 16.13 x 11 inches, p. 89.

Hesse-Honegger, Cornelia. *Four Seed Bugs, Lygaeus equestris*, from Tubre and Santa Maria Madonna, Italy, 1994–1995. Watercolor; 47 x 55 cm. Courtesy of the artist, p. 54.

Höch, Hannah. *Strange Beauty (Fremde Schoenheit)*, 1929. Photomontage with watercolor; 4.43 x 8 inches. Collection: Private, Paris, p. 59.

Höch, Hannah. *Monument I* (DENKMAL I), 1924. Photomontage with watercolor on paper; 11.69 x 6.13 inches, p. 88.

Huan, Zhang. *Family Tree*, 2000. C-print on Fuji archival paper; 50 x 40 inches each (9 prints). Courtesy of the artist, p. 39.

Isaacs, John. *The Matrix of Amnesia*, 1997. Mixed media (wax and resin); 200 x 130 x 60 cm. Collection: Private, p. 63.

Jones, Ronald. *Untitled (Core of the Human Retrovirus: Human T-Lymphotic Virus Type I which contains Protein p25, the RNA which carries the Virus' genetic information, and the Enzyme Reverse Transcriptase, corresponding to the Viral RNA)*, 1989. Bronze, wood, black granite, stainless steel; 63 x 12 x 12 inches overall. Courtesy of the artist and Metro Pictures, New York, p. 190.

Kac, Eduardo. *GFP Bunny*, 2000. Alba, the fluorescent rabbit. Courtesy of the artist and Julia Friedman Gallery, Chicago, p. 94.

Kahlo, Frida. *Henry Ford Hospital*, 1932. Oil on metal; 30.5 x 38 cm. Collection: Museum Dolores Olmeda Patino, p. 117.

Kahlo, Frida. *My Grandparents, My Parents, and I*, 1936. Oil and tempera on metal; 30.7 x 34.5 cm. Courtesy Los Museos Diego Rivera y Frida Kahlo, p. 118.

Katz, Laurel. *The Spotted Merino/Hooked Beetle Sweater Production System*, 1994. Mixed media, 65 x 68 x 68 inches. Courtesy of the artist, p. 171.

Klein, Yves. *Anthropometry of the Blue Period (Anthropométrie de l'epoque bleue)*, 1960. Pure pigment and synthetic resin on paper laid down on canvas; 156.5 x 282.5 cm. Collection: Musee National d'Art Moderne, Centre Georges Pompidou, p. 25.

Korot, Beryl and Reich, Steve. From *Three Tales* (Act 3 *Dolly*), 2002. Video still. Courtesy of the artists, p. 130.

Lang, Gerhard. *Palaenthropical Physiognomy* [detail], 1991/2000. Photographic projections. Courtesy of the artist, p. 100.

le Brun, Charles. *Etude de Taureau et de Boeuf*, 17th Century. 45.5 x 60.5 cm. Collection: Paris Musee de Louvre, p. 12.

le Brun, Charles. *Visage*, 17th Century. 47.5 x 60.5 cm. Collection: Paris Musee de Louvre, p. 12.

Levy, Ellen. *Recombinant + DNA*, 2002. Digital print; 84 x 83.65 inches. Courtesy of the artist, p. 176, 177.

Lindgren, Scott. *Cephalothoracopagus*, 1990. R-print. Courtesy of the artist, p. 56.

Magritte, René. *The Spirit of Geometry (L'Esprit de géométrie)*, 1937. Gouache on paper; 375 x 292 mm. Courtesy Tate, London, p. 59.

Manglano-Ovalle, Iñigo. *Banks in Pink and Blue*, 1999. Installation: mixed media, liquid nitrogen and human semen. Courtesy Max Protetch Gallery, New York, p. 142.

Manglano-Ovalle, Iñigo. *Lu, Jack, and Carrie* (from *Garden of Delights*), 1998. C-print of DNA analyses. Courtesy Max Protetch Gallery, New York, p. 34.

McCarthy, Paul. *Tomato Heads* [detail], 1994. 62 objects, fiberglass, urethane, rubber, metal, clothing; 85 x 54 x 44 inches. Courtesy Luhring Augustine, New York, p. 98.

Mbuti design, Pygmy women of the Ituri Forest. *Zaire drawing*, 20th century. Pigment on bark cloth; 86 x 47 cm. Courtesy Hansjorg Mayer, p. 25.

Michaux, Henri. *Alphabet* [detail], 1927. Ink on paper; 14.25 x 10.25 inches. Collection: Private, p. 21.

Michaux, Henri. *Untitled (Movements [Mouvements])*, 1950–1951. India ink on paper; 12.6 x 9.5 inches. Collection: Private. p. 21.

Miller, Larry. *Genetic Code Copyright Certificate*, 1992. Ink on paper; 8.5 x 11 inches. Courtesy of the artist, p. 174.

Miller, Larry. *Genomic License*, p. 174.

Miller, Steve. *Eat Protein #230*, 2002. Silkscreen and graphite on museum board; 30 x 40 inches. Courtesy of the artist, p. 36.

Miller, Steve. *Eat Protein #259*, 2002. Silkscreen and graphite on museum board; 30 x 40 inches. Courtesy of the artist, p. 36.

Modigliani, Amedeo. Source unknown (possibly destroyed), p. 16.

Moore, Frank. *Cocktail*, 2001. Oil on canvas over featherboard; 15 x 48 inches. Collection: Suzanne Slesin and Michael Steinberg, New York; Courtesy Sperone Westwater, New York, p. 169.

Moore, Frank. *Oz*, 1999–2000. Oil on canvas over featherboard with roadmap on aluminum frame; 70 x 120.5 inches overall. Collection: Howard Stein, New York; Courtesy Sperone Westwater, New York, p. 153.

Moore, Frank. *Wizard*, 1994. Oil on canvas over wood with pharmaceuticals cast in Lucite in an aluminum frame; 68 x 95.5 inches. Collection: Private, Italy; Courtesy Sperone Westwater, New York, p. 165.

Nauman, Bruce, *Ten Heads Circle/Up and Down*, 1990, p. 51.

Nes, Adi. *Untitled (The Featherless Chicken)*, 2002. Color photograph; 60 x 90 cm. Courtesy of the artist and Dvir Gallery, Israel, p. 170.

Oatman, Michael. *Long Shadows: Henry Perkins and the Eugenics Survey of Vermont (Vermont Pure)*, 1995/2000. Installation with video, historical eugenics artifacts and works from permanent collection of the Fleming Museum. Courtesy of the artist, p. 137.

Orlan. *Orlan with Imaginary Generic no. 31: Successful Operation(s)*, 1990, p. 74.

Parker, Olivia. *Forms in Blue*, 1999. Iris inkjet print; 33 x 33 inches. Courtesy of the artist, p. 38.

Pollock, Jackson. *Stenographic Figure*, 1942. Oil on linen; 40 x 56 inches; 101.6 x 142.2 cm. Collection: The Museum of Modern Art, New York, p. 22.

Pondick, Rona. *Untitled Animal*, 1999–2001. Stainless steel; 6.5 x 44.5 x 21.5 inches. Courtesy Sonnabend Gallery, New York, p. 64.

Purcell, Rosamond. *19th century skeleton of hydrocephalic child (Tiny Fetus)*, 1995. Iris print. Courtesy of the artist, p. 70.

Purcell, Rosamond. *Stained and cleared specimen of hydrocephalic child*, 1995. Iris print. Courtesy of the artist, p. 122.

Quinn, Marc. *Lucas*, 2001. Human placenta and umbilical cord, stainless steel, Perspex, refrigeration equipment; 80.5 x 25.19 x 25.19 inches; 204.5 x 64 x 64 cm. Courtesy Jay Jopling/White Cube, London, p. 122.

Quinn, Marc. *Self,* 1991. Blood, stainless steel, perspex and refrigeration equipment; 81.8 x 24.8 inches. Courtesy Jay Jopling/White Cube, London, p. 161.

Quinn, Marc. *Sir John Sulston: A Genomic Portrait,* 2001. Stainless steel, polycarbonate jelly, bacteria colonies, cloned human DNA; approximately 8.5 x 11 inches. Courtesy Jay Jopling/White Cube (London), p. 9.

Ray, Charles. *Family Romance,* 1993. Mixed media; 54 x 96 x 24 inches. Courtesy Donald Young Gallery, Chicago, p. 75.

Redon, Odilon. *The Cyclops,* ca. 1914. Oil on canvas; 64 x 51 cm. Collection: Museum Kroller-Mueller, Otterlo, The Netherlands, p. 58.

Rees, Michael. *Ajna Spine Series 1.11,* 1998. Polycarbonate, selective laser sintering; 20 x 5 inches. Courtesy of the artist, p. 74.

Rockman, Alexis. *Biosphere: Laboratory,* 1993. Oil on two wood panels; 96 x 128 inches. Courtesy Gorney Bravin + Lee, New York, p. 102.

Rockman, Alexis. *The Farm,* 2000. Oil and acrylic on wood; 96 x 120 inches. Collection: The Joy of Giving Something, New York; Courtesy Gorney Bravin + Lee, New York, p. 103.

Rubenstein, Bradley. *Untitled (Boy with Puppy-Dog Eyes),* 1994–1995. Lacquer-based ink on resin coated paper. Courtesy Universal Concepts Unlimited, New York, p. 101.

Ruiz i Altaba, Ariel. *The Three Graces,* 2002. Courtesy of the artist, p. 121.

Rule, Nicolas. *Black Tie Affair,* 1992. Courtesy of the artist, p. 135.

Rupp, Christy. *New Labels for Genetically Engineered Food,* 2000. Vinyl labels on plastic take-out containers; 35 containers. Courtesy of the artist, p. 169.

Schmidt-Rottluff, Karl. *Portrait of Emy,* 1919. Oil on canvas; 73.5 x 65.5 cm. Collection: North Carolina Museum of Art, p. 16.

Schmidt-Rottluff, Karl. *Self-Portrait with Hat,* 1919. Oil on canvas. Framed 85 x 77 cm; Unframed 73.3 x 65 cm. Collection: The Cleveland Museum of Art, p. 16.

Schmidt-Rottluff, Karl. Source unknown, p. 16.

Schneider, Gary. *Chromosomes* (from *Genetic Self Portrait*) [detail/Y chromosome], 1997. Platinum/Palladium print; 5.37 x 3.81 inches [detail]. Courtesy of the artist, p. 31.

Schneider, Jane. *What's That?,* 2002. Wood and mixed media; 128 x 40 x 21 inches. Courtesy of the artist and June Kelly Gallery, New York, p. 99.

Serrano, Andres. *Frozen Sperm II,* 1990. Cibachrome, silicone, plexiglass, wood frame; Unframed 40 x 60 inches; 101.6 x 152.4 cm; Framed 45.13 x 65 inches; 114.6 x 165.1 cm. (Edition of 4). Courtesy Paula Cooper Gallery, New York, p. 141.

Sheehan, Heather. *egg,* 1999. Clay, vaseline, pvc, wood, metal, hose; 45 x 12 x 12 inches; 115 x 30 x 30 cm. Courtesy of the artist, p. 146.

Sherman, Cindy. *Untitled,* 1992. Color photograph; 68 x 45 inches [edition of 6]. Courtesy Metro Pictures, New York, p. 67.

Smith, Kiki. *Untitled (sperm),* 1989–1990. Lead crystal; dimensions variable. Courtesy Pace Wildenstein, New York, p. 139.

Smith, Kiki. *Untitled,* 1989. Gampi paper and methyl cellulose; life size. Courtesy Pace Wildenstein, New York, p. 119.

Storey, Helen and Kate. *Design 11: Anaphase dress* (from *Primitive Streak*), 1997. Courtesy of the artists, p. 162.

Storey, Helen and Kate. *Design 26: DNA sequence/Spinal Column dress in red silk chiffon* (from *Primitive Streak*), 1997. Courtesy of the artists, p. 163.

Sutton, Eva. *Hybrids* [detail], 2000. Installation on LCD flat-panel; dimensions variable. Courtesy of the artist, p. 106.

Tissue Culture and Art Project (Oron Catts, Ionat Zurr and Guy Ben-Ary). *Pigs Wings Project: The Chiropteran Version, the Aves Version, and the Pterosaurs Version,* 2000–2001. Mesenchymal cells (bone marrow stem cells) from pig and biodegradable/ bioabsorbable polymers. Courtesy of the Tissue Culture and Art Project, p. 96.

Trockel, Rosemarie. *Hydrocephalus II*, 1982. Painted plaster and graphite; 18 inches high. Joshua Gessel Collection, p. 69.

Trockel, Rosemarie. *Untitled (Pennsylvania Station)*, 1987. Glass, wood, steel, plaster, hair, synthetic polymer paint, celluloid, cast iron and sheet steel; 43.5 x 192.75 x 48 inches; 120.3 x 490.2 x 121.9 cm. Courtesy Barbara Gladstone Gallery, New York, p. 84.

Velázquez de Silva, Diego. *Las Meninas*, 1656. Oil on canvas; 318 x 276 cm. Collection: Museo del Prado, Madrid, p. 49.

von Hagens, Gunther. Posed whole body specimen with skin. Courtesy Instititut für Plastination, Heidelberg, Germany, p. 159.

von Hagens, Gunther. Reclining woman in the 8th month of pregnancy. Courtesy Institut für Plastination, Heidelberg, Germany, p. 159.

Wagner, Catherine. *-86 Degree Freezers: 12 areas of concern and crisis*, 1995. 12 panel typology; gelatin silver prints; 20 x 24 inches each; 8 x 5 feet installation. Collection: Museum of Fine Art, Houston; Courtesy of the artist, p. 179.

Wagner, Catherine. *Dividing Cell*, 1999. Iris print; 32 x 44 inches. Courtesy of the artist, p. 146.

Warhol, Andy. *Thirty Are Better Than One*, 1963. Synthetic polymer paint and silkscreen ink on canvas; 110 x 94.5 inches, p. 148.

Warhol, Andy. *Untitled (Tattooed Lady)*, 1955. Offset lithograph on paper; 29 x 11 inches, p. 157.

Wilding, Faith. *Embryoworlds* [detail], 1998. Watercolor and ink on vellum; 8 x 6 inches. Courtesy of the artist, p. 143.

Witkin, Joel-Peter. *Portrait of a Dwarf*, 1987. Toned gelatin silver print; 16 x 20 inches. Courtesy Fraenkel Gallery, San Francisco, p. 72.

Witkin, Joel-Peter. *Portrait of the Holocaust*, 1982. Toned gelatin silver print; 16 x 20 inches. Courtesy Fraenkel Gallery, San Francisco, p. 120.

Xiaogang, Zhang. *Bloodline: The Big Family, Number 2*, 1995. Courtesy of the artist and Chinese Contemporary, London, p. 133.

Xiaogang, Zhang. *Bloodline: Portrait of a Family*, 1993. Courtesy of the artist and Chinese Contemporary, London, p. 133.

Group Art Exhibits That Focus on Genetics Concepts

This list of group exhibitions illustrates the increasing trend for university art galleries, museums, and cultural institutions to focus on artists' views of the complex topic of genetics, including eugenics, genomics, molecular genetics, and biotechnological applications, especially as this science relates to issues of ethics and society. We have included only group exhibits for the sake of brevity and to illustrate the trend, but we recognize that there have been many excellent single-artist exhibits throughout the world that address these topics. We have also excluded exhibits at science museums that focus essentially on the science.

1992 *Post-Human,* FAE Musee d'Art Contemporain, Lausanne; Castello di Rivoli, Museo d'Arte Contemporanea, Rivoli; Deste Foundation for Contemporary Art, Athens. Curated by Jeffrey Deitch.

1993 *Genetic Art,* Ars Electronica, Linz, Austria. Organized by Peter Weibel.

1994 *Gene Culture: Molecular Metaphor in Visual Art.* Plaza Gallery, Fordham College at Lincoln Center, New York, New York. Curated by Suzanne Anker.

1998 *Art + Bio.* Central Michigan University. Curated by David Stairs.

Out of Sight: Imaging/Imagining Science. Santa Barbara Museum of Art. Curated by Karen Sinsheimer.

Gene Worlds, Bundeskunsthalle, Bonn, Germany. Curated by Edith Decker.

1999 *Life Sciences, Ars Electronica,* Linz, Austria. Organized by Gerfried Stocker and Christine Schopf.

2000 *Tenacity: Cultural Practices in the Age of Information and Biotechnology.* Swiss Institute, New York, New York. Curated by Yvonne Volkart.

Paradise Now: Picturing the Genetic Revolution. Exit Art, New York, New York; University of Michigan Museum of Art, University of Michigan, Ann Arbor; Tang Teaching Museum, Skidmore College, Saratoga Springs, New York; plus additional venues. Co-curated by Marvin Heiferman and Carole Kismaric.

Unnatural Science: An Exhibition, MASS MoCA, North Adams, Massachusetts. Curated by Laura Steward Heon.

Spectacular Bodies: The Art and Science of the Human Body from Leonardo to Now, Hayward Gallery, London. Curated by Martin Kemp and Marina Wallace.

Working Drafts, Two Ten Gallery of the Wellcome Trust, London, England. Curated by Denna Jones.

2001 *Perfecting Mankind: Eugenics and Photography*. International Center of Photography, New York, New York. Curated by Carol Squiers.

Gene(sis) Contemporary Art Explores Human Genomics. Henry Art Gallery, University of Washington, Seattle. Travels to Berkeley Art Museum, Frederick Weisman Museum of Art, The Mary and Leah Block Museum of Art. Curated by Robin Held.

Under the Skin: Transformations in Contemporary Art, Stiftung Wilhem Museum, Germany. Curated by Renate Heidt Heller.

2002 *Foreign Body: Photography and the Prelude to Genetic Modification*. International Center of Photography, New York, New York. Curated by Carol Squiers.

Converge: Where Art and Science Meet, Adelaide Biennial, Art Gallery of South Australia.

PhotoGENEsis Artists' Response to the Genetic Information Age, Santa Barbara Museum of Art. Curated by Karen Sinsheimer.

The *Genius of Genetics: A Celebration of Gregor Mendel through Science and Art*, Abbey of St.Thomas, Brno, The Czech Republic. Curated by Caterina Albano and Marina Wallace.

DNArt, Biennale Kunst Meran, Italy. Curated by Valerio Dehò.

2003 *Genetic Expressions: Art after DNA*, Heckscher Museum of Art, Huntington, New York. Curated by Lynn Gamwell and Elisabeth Merryman.

Codes and Identities, Clifford Art Gallery, Colgate University. Curated by Linn Underhill.

Brave New World, Organization of Independent Artists, New York, New York. Curated by Amy Wilson.

Women in Science (Genomically Yours), Universal Concepts Unlimited, New York, New York.

From Code to Commodity: Genetics and Visual Art, New York Academy of Sciences Art Gallery. Curated by Suzanne Anker and Dorothy Nelkin.

Genomic Issue(s): Art and Science, Art Gallery of the Graduate Center, City University of New York. Curated by Marvin Hieferman.

How Human: Life in the Post-Genome Era, International Center for Photography, New York, New York. Curated by Carol Squiers.

L' Art Biotech, Le Lieu Unique, Nantes, France. Curated by Jens Hauser. *Four Plus: Writing DNA*, Wellcome Trust, London. Curated by Denna Jones.

Index